CATALYSIS IN PETROCHEMICAL PROCESSES

CATALYSIS IN PETROCHEMICAL PROCESSES

SAMI MATAR
MANFRED J. MIRBACH
HASSAN A. TAYIM

KLUWER ACADEMIC PUBLISHERS
DORDRECHT / BOSTON / LONDON

Library of Congress Cataloging in Publication Data

Matar, Sami, 1931-
Catalysis in petrochemical processes / Sami Matar, Manfred J. Mirbach, Hassan A. Tayim.
p. cm.
Includes index.
ISBN 978-90-277-2721-3
1. Catalysis. 2. Petroleum chemicals. I. Mirbach, Manfred J. II. Tayim, Hassan A. III. Title.
TP156.C35M38 1988
661'.804--dc19 88-4391
CIP

Published by Kluwer Academic Publishers,
P.O. Box 17, 3300 AA Dordrecht, The Netherlands.

Kluwer Academic Publishers incorporates the publishing programmes of
D. Reidel, Martinus Nijhoff, Dr W. Junk, and MTP Press.

Sold and Distributed in the U.S.A. and Canada
by Kluwer Academic Publishers,
101 Philip Drive, Norwell, MA 02061, U.S.A.

In all other countries, sold and distributed
by Kluwer Academic Publishers Group,
P.O. Box 322, 3300 AH Dordrecht, The Netherlands.

Table of Contents

Preface

The field of petrochemicals started some years ago with the simple addition reaction of water to propylene for the production of isopropyl alcohol. Currently, the petrochemical industry has become a multi-billion dollar enterprise which encompasses a wide field of chemical products. Almost all the basic organic reactions such as hydrogenation, alkylation, substitution, polymerization, etc. are utilized for the production of these chemicals. It may not, however, have been possible to establish this huge industry without the use of different catalysts. In other words, the great advancements in the catalytic area have supported the vast developments in the petrochemical field.

In this book, we have adopted the idea of discussing the petrochemical industry from the point of view of reactants' activities and susceptibilities toward different catalysts. The book is thus classified according to the reaction type. This will enable students and other users of the book to base their understanding of the petrochemical field on the fundamental principles learned in chemistry. However, the first chapter is aimed at establishing some basic facts on the petrochemical industry and its major uses. It discusses, without going into details, the raw materials used, the intermediates and the downstream products. The next eight chapters discuss in some detail the main reactions and the catalysts used for the production of chemicals and polymers from petroleum. The last chapter is devoted to a discussion of some of the practical techniques used in the catalytic field.

The book is considered to be an important supplement to the background of students studying industrial chemistry, chemical engineering, and to technical staff working in the petroleum and the petrochemical field. However, the book is intended to be used as a research monograph. The authors felt that a full literature review of the subjects discussed is not necessary. Only selected up-to-date references, which have either been used in compiling the background for the text, or are recommended for supplementary reading, are presented at the end of each chapter. Instructors in the fields of chemistry, chemical engineering and industrial chemistry will find this book a feasible core for a senior undergraduate or a graduate course in applied chemistry.

We, the authors, hope that the state of the art, as it is now, has been well-summarized, fulfilling the objective of publishing this book.

SAMI MATAR
MANFRED MIRBACH
HASSAN TAYIM

CHAPTER 1

Raw Materials and Chemicals from Carbonaceous Sources

1. Introduction

Petrochemicals, or petroleum chemicals, are organic compounds derived from natural gas, associated gas or crude oils. These do not include organic compounds produced from other natural sources, whether from animal or plant origin such as fats, oils, cellulosics etc.

However, petrochemicals may also be derived from such other carbonaceous materials as coal, peat, tar sands and oil shale. Examples of downstream polymers and chemicals from petroleum and other cabonaceous sources include synthetic fibers, plastic rubber, nitrogen fertilizers such as urea, solvents such as ethylene glycol, surface active agents such as synthetic detergents and many others. The petrochemical field is a very diversified one, and several intermediates and chemicals from petroleum are used in specialized organic industries such as pharmaceuticals, pesticides, explosives, fragrances etc. For example, aspirin (acetyl-salicylic acid) is produced by reacting salicylic acid and acetic anhydride.

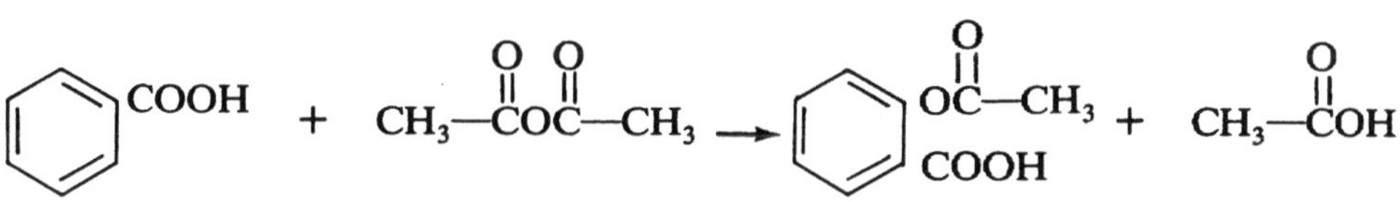

Salicylic acid is produced from phenol and carbon dioxide. Acetic anhydride is produced from methanol and carbon dioxide. All these chemicals are either products or by-products of the petroleum and petrochemical industries. Another example, which illustrates the relation between the petrochemical field and the perfumery field, is the synthesis of phenlethyl alcohol, which has the smell of orange flowers. This alcohol can be prepared by reacting phenyl magnesium bromide (Grignard reagent) with ethylene oxide. It can also be produced by a direct alkylation of benzene with ethylene oxide.

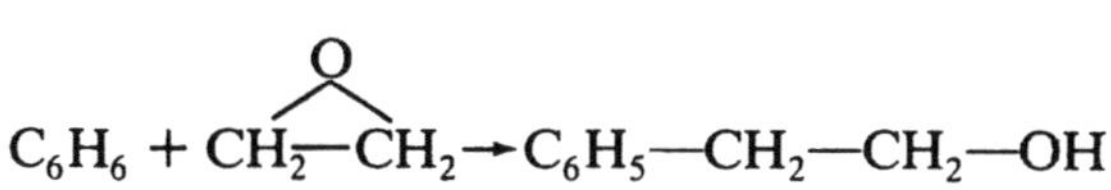

Both benzene and ethylene oxide are important petrochemical intermediates for preparing a substantial number of compounds.

1.1. Raw Materials

As mentioned before, the raw materials used for producing intermediates for the petrochemical industries are natural gas, associated gas, and crude oils. Other carbonaceous materials could also be utilized for the same purpose.

1.1.1. NATURAL AND ASSOCIATED GASES

Natural gas is a term used for a gas mixture found in porous underground reservoirs. If the gas is dissolved under pressure with crude oils, it is called associated gas. Both natural and associated gas are composed of a mixture of light hydrocarbons plus other non-hydrocarbon gases. Depending on the topographical history of each reservoir, the ratio of the gas constituents may differ greatly from one gas reservoir to another. The main constituent of these gases, however, is methane. A major difference between natural and associated gas is the ethane/methane ratio which is usually much higher in associated gas than in natural gas. Table 1-1 shows a typical analysis of some selected natural and associated gases. As can be seen, natural gas may contain a high percentage of other non-hydrocarbon gases such as carbon dioxide, hydrogen sulfide, nitrogen and helium. Both hydrogen sulfide and carbon dioxide are acidic gases which should be eliminated or reduced to non-harmful levels before utilizing the gas. Natural gas is mainly used as a fuel. However, many petrochemical industries are based on natural and associated gas. For example, Saudi Arabian petrochemical plants at Jubail and Yanbu are mainly based on associated gas as a petrochemical feedstock as well as a fuel. Methane, the major component of natural and associated gas, is the simplest hydrocarbon. As a chemical compound it is utilized for the production of synthesis gas. Synthesis gas is the precursor of many important chemicals.

Natural Gas Treatment

Before utilizing natural gas as a fuel or as a petrochemical raw material, it has to be treated to remove or reduce the amount of acid gases, adjust the humidity and free the gas from heavier hydrocarbon components.

Table 1-1. Typical analysis of selected natural and associated gases.

	Natural Gases			Associated Gases		
Component	Salt Lake U.S.	Kliffside U.S.	Larcq France	Libya	Saudi Arabia Abqaiq	North Sea U.K.
Methane	95.0	65.8	70.0	66.8	62.24	85.9
Ethane	0.8	3.8	3.0	19.4	15.07	8.1
Propane	0.2	1.7	1.4	9.1	6.64	2.7
Butane	...	0.8	0.6	3.5	2.40	0.9
Pentanes & heavier	...	0.5	...	1.2	1.12	0.3
Hydrogen sulfide	...	...	15.0	...	2.80	...
Carbon dioxide	3.6	0.6	10.0	...	9.20	1.6
Nitrogen	0.4	25.6	...	...	...	0.5
Helium	...	1.8	...	...	...	...

Hydrogen Sulfide and Carbon Dioxide Removal

Treatment of acid gases found in natural or associated gases can be effected either by a chemical absorption scheme or by the use of physical absorbents. In the first scheme, natural gas is passed through an easily regenerated base such as diethanolamine. The solvent used, which contains carbonates and sulfides, is then regenerated by steam treatment. Other amines are also used for the same purpose such as monoethanolamine and diglycolamine. The aqueous solution of the latter has a lower freezing point than the other amines which makes it usable under colder conditions. It also has a high treatment capacity at elevated temperatures.

Physical absorbents are compounds which have the ability to dissolve the acid gases without forming chemical bonds with them. Physical absorbents are more commonly used when the gas contains a high percentage of carbon dioxide. In the Selexol process, for example, the physical solvent is the dimethylether of polyethylene glycol (CH_3—$[OCH_2CH_2]_n$—OCH_3).

These solvents are more active at higher pressures and they are therefore suitable for treating compressed natural gas. Solid adsorbents are also used for the treatment of acid gases. Molecular sieves (aluminosilicate compounds) are solid compounds of this type which may also be utilized for adjusting the gas humidity by adsorbing water.

Humidity Treatment

Natural and associated gases have to be dehydrated to prevent hydrate formation during the transfer of the gas by pipeline. Hydrates are solid compounds formed by a physico-chemical reaction between the hydrocarbon gases and water under high pressures.

Treatment starts by passing the gas under normal pressure through a polar solvent which dissolves water but not the hydrocarbon gases. Solvents used for this purpose should have a low vapor pressure, such as ethylene glycol (EG), diethylene glycol (DEG) and triethylene glycol (TEG). The last solvent is preferred in vapor-phase processes because of its low vapor pressure which results in low glycol losses. Molecular sieves, as mentioned above, could be used for the dual purpose of reducing the humidity as well as removing the acid gases from natural and associated gases.

Heavy Hydrocarbons Removal

After treating the gas as described above, it is now dry and sweet. A further step has to be undergone before using the gas, and that is the removal of gases heavier than methane by a cooling scheme. Liquid propane is used to cool natural gas. The cold gas is then washed with a hydrocarbon liquid which dissolves the condensable hydrocarbons. Methane mixed with a small amount of ethane is collected for use as a fuel or as a petrochemical feedstock. Table 1-2 is a typical analysis of a treated natural gas. The condensable hydrocarbons normally known as 'Natural Gas Liquids', are separated from the hydrocarbon solvent and then fractionated.

Table 1-2. Typical analysis of a treated natural gas.

Component Mole %	Before Treatment	After Treatment
C1	70.35	94.85
C2	0.83	0.99
C3	0.22	0.003
C4	0.13	0.004
$C5^+$	0.17	0.018
N_2	0.45	0.62
CO_2	27.85	3.50
H_2S	0.0013	—

Ethane may be separated from the upper fractionation zone and used as a feed to an ethane cracker. Both propane and butane are usually separated together, then liquefied and used as liquefied petroleum gas (LPG). The last fraction, composed of hydrocarbons heavier than C-4, is called 'Natural Gasoline'. This fraction may contain some unseparated butanes and is customarily added to gasoline to raise its vapor pressure.

1.1.2. CRUDE OILS

A crude oil is a black viscous mixture which varies greatly in density from one oil well to another. Unlike natural gas, a crude oil is mainly composed of a heavier mixture of hydrocarbons. The principle hydrocarbon classes of the mixture are; paraffins, cycloparaffins (naphthenes) and aromatics. Olefinic compounds are essentially not present in crude oils. Depending on the origin of the crude oil and on the conditions under which it is formed, the ratio of these hydrocarbon classes varies appreciably from one crude to another. In addition to the hydrocarbons, crude oils may contain appreciable amounts of sulfur, nitrogen and oxygenated hydrocarbon derivatives, along with smaller amounts of organometallic compounds. Iron, nickel and vanadium compounds are examples of organometallics which may be found in trace amounts in crude oils and their fractions. However, the presence of nickel and vanadium compounds above certain levels in oil fractions and fuel oils is poisonous to many catalysts used in petroleum and petrochemical processes. Table 1-3 shows a typical analysis of some representative crude oils.

Crude oils, are generally classified into three types depending on the ratio of the hydrocarbon class that predominates in the mixture. These are:

a — Paraffinic: — constituents are predominantly paraffinic hydrocarbons. Low percentage of aromatics and asphaltenes.

b — Naphthenic: — contain relatively a higher ratio of cycloparaffins, and a higher amount of asphalt than in paraffinic crudes.

Table 1-3. Typical analysis of some representative crude oils.

Tests	Jalo (Libya)	Alamein (Egypt)	Amer (Egypt)
Specific Gravity @ 60/60°F		0.8576	0.9255
Gravity API @ 60°F	33.8	33.41	21.39
Water Content vol %	0.05	0.2	0.1
Salt Content wt %	0.003	0.007	0.005
Total Sulfur wt %	0.68	0.86	4.5
Hydrogen Sulfide, ppm	—	—	6311
Mercaptan Sulfur, ppm	—	—	86
Pour Point, °F	9	35	30
Conradson Carbon Residue wt %	1.4	4.9	4.5
Ash Content, wt %	0.0012	0.004	0.05
Asphaltene Content wt %	0.25	3.37	8.06
Paraffin Wax Content wt %	4.0	3.3	1.35
Vanadium Content, ppm	9	15	67
Phenol Content, wt %	—	—	0.0082

c — Asphaltic: — contain a relatively large amount of fused aromatic rings and a high percentage of asphalt.

However, there is no sharp borderline separating these three classes. In general, paraffinic crudes are less dense and are easier to process than the other two classes.

The specific gravity (or API) is an important parameter which is sometimes used to indicate the crude-oil quality. A low density crude (high API) has a higher percentage of the more valuable light and middle distillates. Oils with a high sulfur content are less in demand than those with a low sulfur content since fractions from the former oils have to be subjected to further treatment to remove these sulfur compounds (see Chapter 3, 'Hydrodesulfurization').

In processing crude oils, naphtha is obtained from the upper fractionator zone. Naphtha is composed of a light hydrocarbon mixture approximately in the range of C-5—C-11.

Naphtha obtained from a highly naphthenic crude is easier to catalytically reform to a high octane gasoline than naphtha from a paraffinic crude, since the rate of aromatization of naphthenes is faster than that of paraffins. Catalytic reforming is a process used to change cycloparaffins and paraffins in the naphtha range to aromatics (see Chapter 3, 'Dehydrogenation'). It is a major process for producing benzene, toluene and xylenes (BTX). These compounds are precursors to many other chemicals. Middle fractions obtained from distilling crude oils are kerosine and gas oil. Fuel oil is taken from the lower zone of the fractionator. These fractions are mainly used as fuels for different purposes. However, gas oils and fuel oils may be further catalytically cracked to produce more gasoline and kerosine. The gasoline obtained from catalytic crackers is highly aromatic and may also be used to separate BTX compounds for petrochemical use. Kerosine is mainly used as a jet fuel. However, *n*-paraffins for detergent production may be separated from kerosines.

1.1.3. COAL, PEAT, OIL SHALE AND TAR SANDS

Coal, peat, oil shale and tar sands are materials which are mainly constituted of carbon. The H/C ratio of these materials is appreciably lower than that of most crude oils, (Table 1-4). They are classified, however, with crude oils as carbonaceous materials. Their origin may bear some similarity to that of crude oils. In a way, these substances were formed by the action of heat and pressure on some organic remains — whether of animal or plant origin — which decomposed to the final form (oil, coal, etc.). It is generally accepted, for example, that coal originated from land plant debris that decayed under mildly reducing conditions, while crude oils are formed from sea plants and animals decaying under highly reducing conditions. Peat is considered to be immature or young coal.

Coal and Peat

Coal is one of the major fossil energy sources of our time. It has been responsible for powering the industrial revolution in Europe and America for more than a century. However, since World War II, coal as a primary energy source has been overshadowed by oil, which is a cleaner, more convenient fuel to handle and process. However, coal is currently supplying about 28% of the world's energy needs.

Coal has been and still is a source of organic chemicals. For example, liquid coke-oven by-products from coal carbonization still provide about 6% of the benzene demand. If coal gasification and liquefaction become economically feasible in the future, coal could be a major source of organic chemicals.

Coals are classified into different ranks according to the degree of chemical change that has taken place due to the decay of the remains. Peat, sub-bituminous, bituminous and anthrathite are different ranks of coal. The first type being the one that has been subjected to the least change, while the last has suffered the highest coversion. Anthracite is characterized by having a higher C/H ratio than other coal ranks. Table 1-4 shows the carbon, hydrogen and oxygen analysis of some coals, compared with a typical crude oil.

Table 1-4. Typical elemental analysis of some coals compared with a crude oil.

	C	H	S	N	O	H/C Ratio
Crude Oil	84.6	12.8	1.5	0.4	0.5	1.82
Peat	56.8	5.6	0.3	2.7	34.6	1.18
Lignite	68.8	4.9	0.7	1.1	24.5	0.86
Bitumenous Coal	81.8	5.6	1.5	1.4	9.7	0.82
Anthrathite	91.7	3.5			2.7	0.46

The higher C/H ratio of coals in comparison to crude oils may be a manifestation of the strong reducing conditions under which crude oils have been formed. Adjustment of the C/H ratio of coals, in order to be comparable to that of crude oils, may be done through liquefaction or catalytic hydro-liquefaction processes. Products from these conversion processes are called synthetic crude oils, which may further be treated to produce conventional fuels and chemicals.

Oil Shale

Oil shale is a naturally low permeability rock made of an inorganic mineral matter which is interspersed with a high molecular-weight organic waxy substance called 'kerogen'. This waxy substance is not very soluble in organic solvents. When shale is heated, kerogen decomposes to a liquid-shale oil that can be processed to products similar to petroleum fractions. The oil yield ranges between 10—20 wt % of the total shale.

Oil shale composition may differ substantially from one shale to another. Table 1-5 shows the analysis of two shale deposits (Eastern and Western) in U.S. The 'Fischer assay', is a test that indicates the amount of kerogen that can be obtained from the shale.

To separate kerogen from the inorganic matter, shale is subjected to a retorting process in which crushed shale is heated to high temperatures to pyrolyze the high molecular weight organic polymer. Oil produced by retorting shales is a thick viscous material which is sometimes called synthetic crude oil.

The H/C ratio (approximately 1.6) of this oil is somewhat lower than that of most crude oils. However, its composition is quite different. The nitrogen and the oxygen contents are usually higher but the sulfur content is much lower than that of crude oils. Attempts to utilize the organic matter of the huge shale deposits through retorting followed by hydrotreatment have been successful. However, an obstacle that hinders the progress in this line, is the disposal of the massive solids from retorting processes.

Table 1-5. Typical analysis of U.S. eastern and western shales.

	Eastern	Western
Ultimate Analysis, wt % (dry basis)		
Organic Carbon	13.7	13.6
Hydrogen	1.64	2.1
Sulfur	4.72	0.5
Carbon Dioxide	0.5	15.9
Ash	78.3	66.8
Fisher Assay Analysis		
Oil Yield, wt %	4.6	11.4
Water Yield, wt %	2.3	1.6
Loss + Gas, wt %	2.4	2.6
Assay, gal/ton	10.3	29.8

Tar Sands

Tar sands, also called oil sands or bituminous sands, are mainly composed of sand saturated with bitumen and water. Tar sand deposits are found in large accumulations in a few parts of the world. About 98% of all tar sand is found in seven deposits, the largest of which are the Athabasca deposits in Western Canada. These deposits are covered by a semifloating mass of partially decayed vegetation approximately 6 meters thick.

The bitumen, a heavy viscous mixture, is mainly constituted of heavy hydrocarbon compounds. Recovering the bitumen differs from getting crudes from oil wells. Oil sands are either strip mined, if the deposits are near the surface, or recovered *in situ* if they are in deeper beds. Bitumen is then extracted by hot water. The bitumen changes to a fluid when heated and floats on the water surface. Cooling changes it to a semi-solid, viscous, dense material which is heavier than water. Table 1-6 shows a typical analysis of recovered bitumen from Athabasca-deposits. The H/C ratio of tar sands bitumen is intermediate between those of coals and crude oils. Complex polynuclear aromatics are the major components of these bitumens. However, the quality of products obtained by distilling the bitumen could be improved greatly by hydroprocessing schemes.

1.2. Intermediates for Petrochemical Production

Intermediates used for producing different petrochemicals are simple reactive compounds such as olefins, diolefins, alcohols, aldehydes, aromatics etc. These

Table 1-6. Typical analysis of Athabasca bitumen.

API Gravity	6.0
Pour Point °F	11.2
Calorific Value Btu/lb	17,900
Components wt %	
Asphaltenes	20.0
Resins	25.0
Oils	55.0
Ultimate Analysis	
Carbon	83.6
Hydrogen	10.3
Sulfur	5.5
Nitrogen	0.4
Oxygen	0.2
H/C Atom Ratio	1.5
Heavy Metals ppm	
Nickel	100
Vanadium	250
Copper	5

compounds can react with one another to produce downstream chemicals and polymers. Among the most important of these compounds, ethylene is used for producing many chemicals in addition to polyethylene polymers. Propene, butenes, butadiene and isoprene are among the most important olefinic hydrocarbons used for petrochemicals production. Benzene, toluene, and the xylenes (BTX) are precursors of a whole series of chemicals and polymers. Synthesis gas, obtained from natural gas or other carbonaceous sources, is also an important intermediate for many chemicals, from fertilizers to methanol to single-cell proteins.

1.2.1. ETHYLENE, PROPENE AND BUTENES

Ethylene is produced by the thermal, non-catalytic cracking of ethane. Theoretically any hydrocarbon compound could be used as a feed to a cracking unit. A higher yield of ethylene is obtained, however, by cracking ethane than by cracking heavier hydrocarbons. The most important sources for ethylene are ethane, naphtha and gas oil. Naphtha is a light hydrocarbon mixture having a carbon range of approximately C-5—C-11.

Gas oil is an intermediate oil fraction from atmospheric distillation of crude oils. Vacuum gas oil (VGO) is obtained by distilling fuel oil under reduced pressure.

Table 1-7 shows the yield of ethylene from cracking different feedstocks. Ethylene, the first member of the alkene series, is a very reactive compound. Many reagents such as water, chlorine, hydrogen chloride, oxygen, etc., can add to ethylene producing valuable chemicals. It can readily be polymerized or copolymerized to produce polyethylene polymers and copolymers. Fig. 1-1 shows the most important chemicals based on ethylene.

Propene (propylene) is the second member of the alkene series and is also a very reactive hydrocarbon. Similar to ethylene, simple addition compounds can react with propene giving valuable chemicals. Commercial polymers and copoly-

Table 1-7. Typical yields from cracking various feedstocks with ethane recyle.

Feedstock	Products wt %				
	Ethylene	Propylene	Butadiene	BTX	Other
Ethane	84.0	1.4	1.4	0.4	12.8
Propane	44.0	15.6	3.4	2.8	34.2
n-Butane	44.4	17.3	4.0	3.4	30.9
Light Naphtha	40.3	15.8	4.9	4.8	34.2
Full Range Naphtha	31.7	13.0	4.7	13.7	36.9
Light Gas Oil	28.3	13.5	4.8	10.9	42.5
Heavy Gas Oil	25.0	12.4	4.8	11.2	46.6
Crude Resid	21.0	7.0	2.0	11.0	59.0
Crude Oil	32.8	4.4	3.0	14.4	45.4

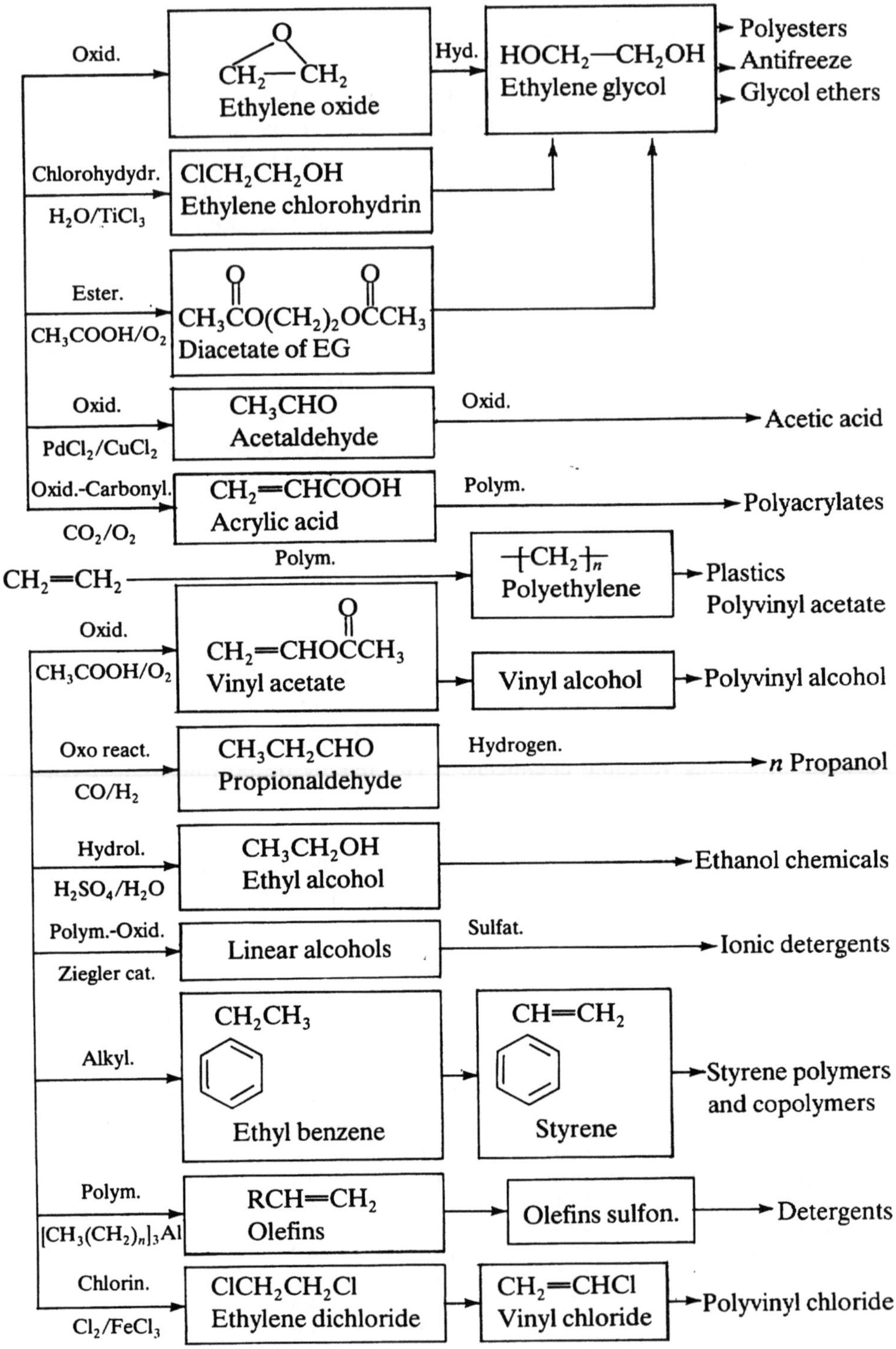

Fig. 1-1. Most important chemicals based on ethylene.

mers of propene are well-known plastics and fibers. Not as many chemicals, however, are produced from propene as from ethylene. This is probably due to the fact that propylene is always a by-product not a main product. There is no special production process for propylene and it always comes as a by-product from different ethylene production lines and from refinery operations. Fig. 1-2 shows the important chemicals based on propylene.

Butenes and butadiene are also by-products of refinery processes and of ethylene production lines. Except for isobutene, few chemicals are produced from C-4 olefins.

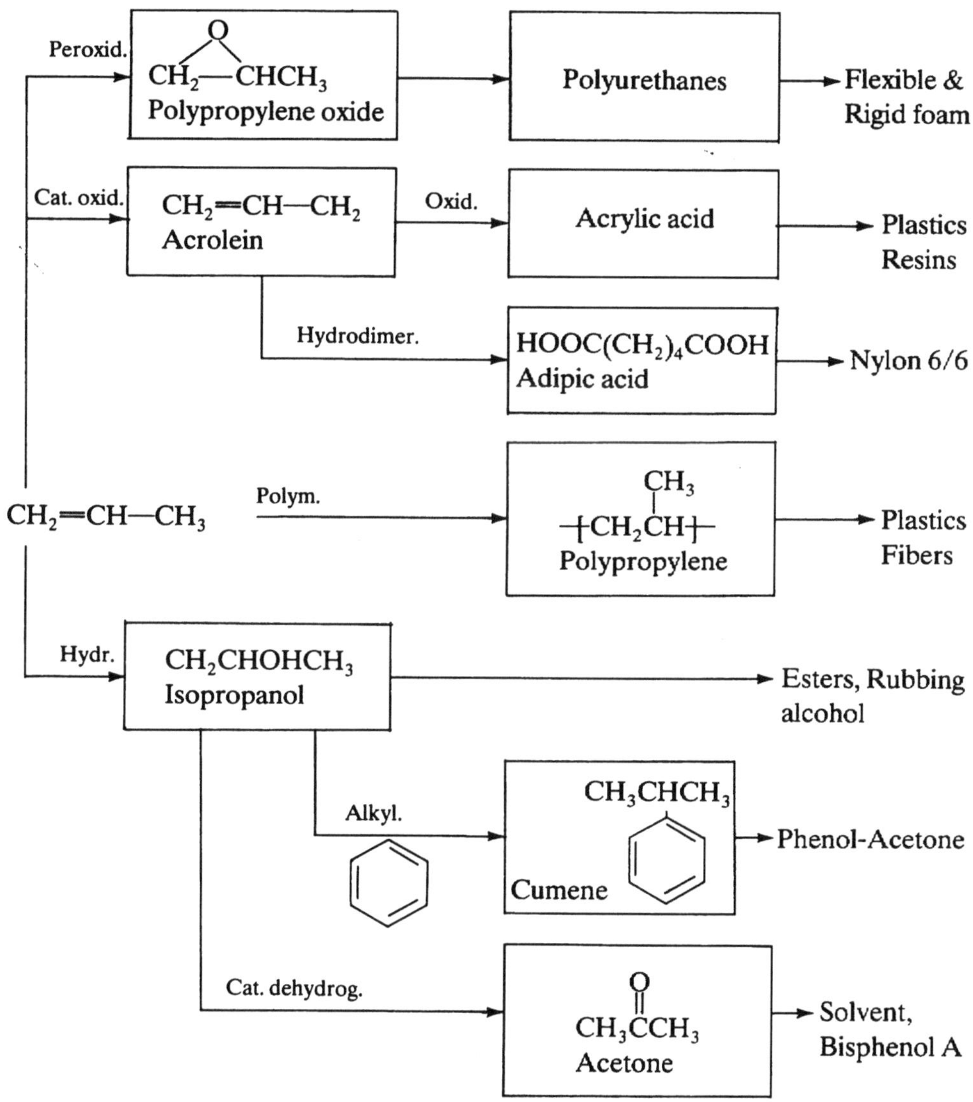

Fig. 1-2. Most important intermediates and chemicals from propylene.

Important chemicals obtained from isobutene by addition of the appropriate reagent are methyl *tert*-butyl ether, *tert*-butyl alcohol and isobutene oxide. Isobutene can be dimerized to diisobutylene and copolymerized with isoprene to butyl rubber. It can be oxidized to methacrylic acid.

1.2.2. BUTADIENE, ISOPRENE AND CHLOROPRENE

These three conjugated dienes are important elastomer precursors. Many of the reactions of conjugated diolefins are similar to those of compounds having isolated double bonds such as ethylene, propene and butenes. However, conjugation results in two types of additions, a 1,2- and a 1,4-addition reaction. For example, butadiene can react with chlorine giving 3,4-dichloro-1-butene by 1,2-addition and 1,4-dichloro-2-butene by 1,4-addition reaction.

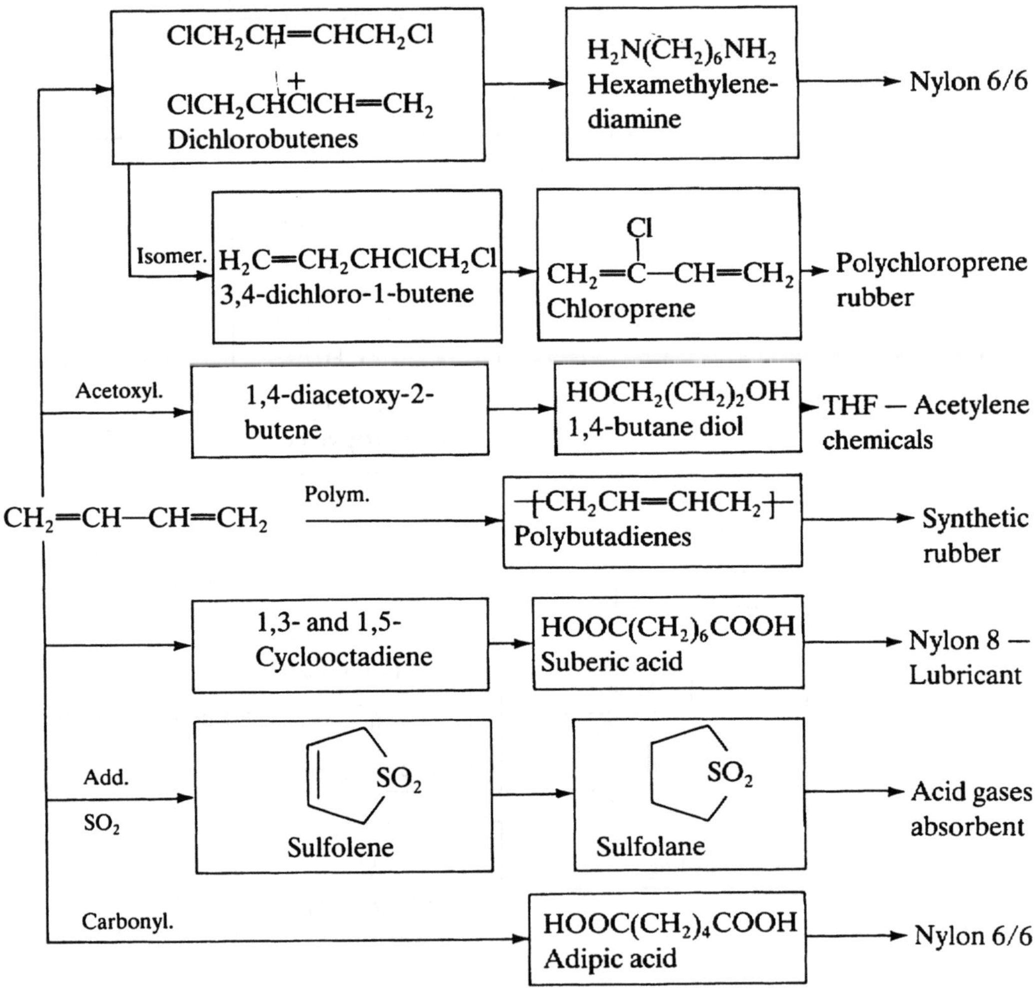

Fig. 1-3. Most important intermediates from butadiene.

Butadiene is commercially produced by cracking naphtha and other hydrocarbons, dehydrogenation of C-4 hydrocarbon mixtures and as a by-product of ethane cracking. It is mainly used for the production of polymers and copolymers. Only a small percentage of butadiene demand is utilized for the production of chemicals. However, some important chemical intermediates are obtained from butadiene (Fig. 1-3). (Stereospecific polymers and copolymers of butadiene are discussed in Chapter 9).

Isoprene (2-methyl-1,3-butadiene) is an important compound for producing polymers which have a similar structure to natural rubber. It can be produced by dehydrogenating a mixture of pentenes from refinery streams and from ethane cracking. Other important methods for isoprene production include the cracking of propene dimers and the cracking of dimethyl dioxane which is synthesized from isobutene and formaldehyde. Polymers with a high *cis*-structure have good mechanical properties and are produced by the use of coordination catalysts (see Chapter 9).

Chloroprene (2-chloro-1,3-butadiene) is also used for producing rubber with excellent chemical resistance. Chloroprene can be produced by chlorination of butadiene. The chlorinated mixture is then isomerized to 3,4-dichloro-1-butene which is dehydrochlorinated to chloroprene.

$$CH_2{=}CH{-}CHCl{-}CH_2Cl \rightarrow CH_2{=}CH{-}CCl{=}CH_2 + HCl$$

1.2.3. BENZENE, TOLUENE AND XYLENES (BTX)

Benzene, toluene and xylenes are mainly produced from catalytic naphtha reforming. Another source of BTX is catalytic cracking, in which heavier, less valuable crude oil fractions and residues are catalytically decomposed to lighter hydrocarbon components. The product naphtha has a high octane number due to aromatization reactions taking place along with cracking. A new catalytic process, the Cyclar process, can be an important source of BTX. The feed is a liquified petroleum gas (mainly C-3 and C-4). The catalyst, which is a zeolite type, has the capability of dehydrogenating the feed, polymerizing the product to unsaturated oligomers, and dehydrocyclizing the oligomers to aromatics. The benzene yield from a Cyclar process is generally higher than from catalytic reforming processes.

Another, less important source for BTX is pyrolysis gasoline from naphtha cracking for ethylene production.

Toluene, which is not used for chemical synthesis, is an additional source of benzene by a hydrodealkylation process.

Benzene Chemicals

Benzene is the most widely used aromatic. Many chemicals and intermediates are derived from it. As an active compound, benzene can be alkylated by many alkylating agents. It can be chlorinated, nitrated, oxidized and hydrogenated. It is not polymerized, however. Derivatives of benzene, such as styrene, are polymerized to useful products. About 50% of chemical benzene is used for producing ethylbenzene, which is mainly utilized to obtain styrene. Fig. 1-4 shows the most important chemicals based on benzene.

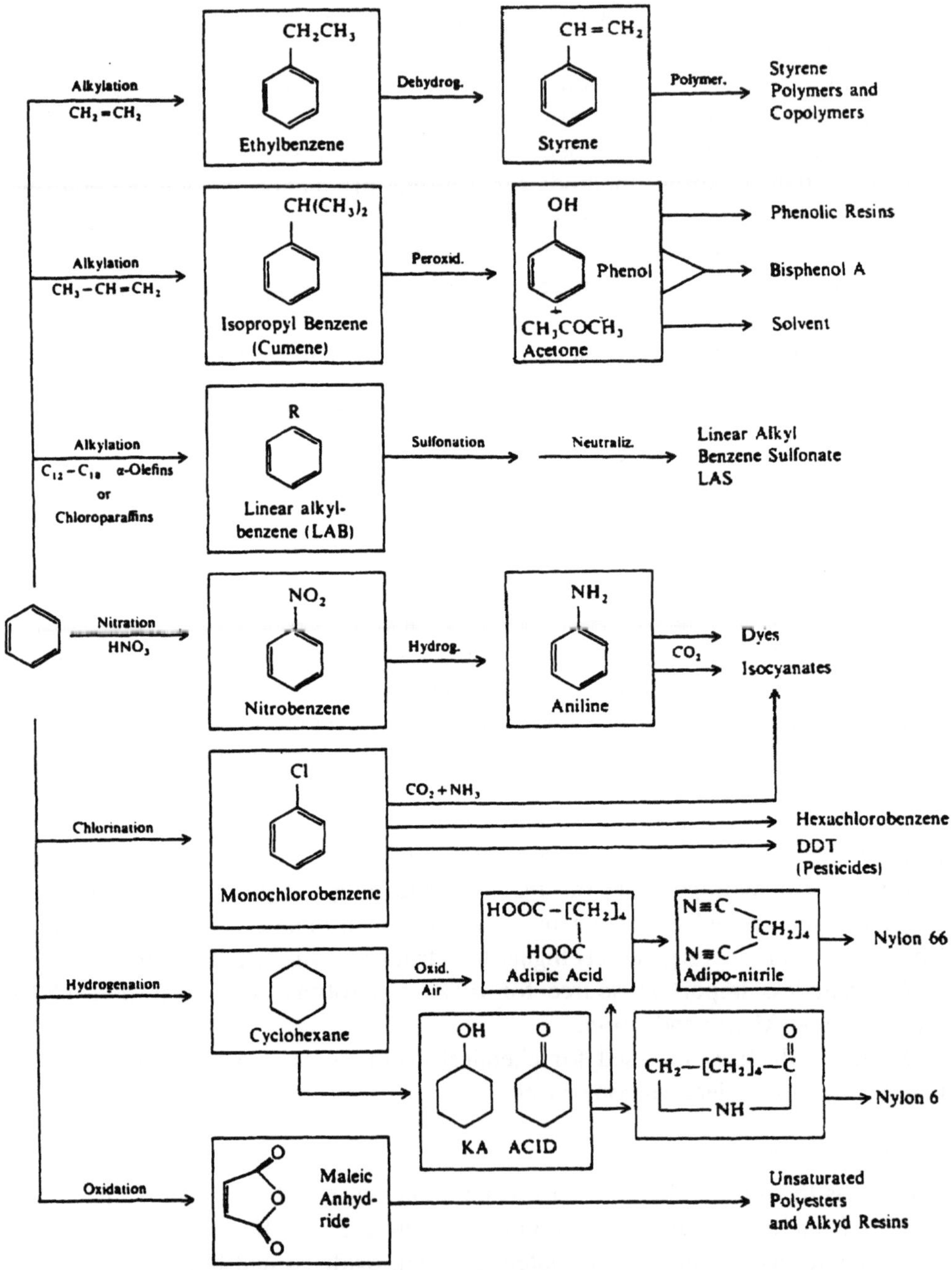

Fig. 1-4. Important intermediates and chemicals from benzene.

Toluene Chemicals

Although toluene is available in much larger amounts than benzene in the reformate gasoline, its use is limited to fewer products. The petrochemical use of toluene represents less than 5% of the total product in the reformate. Almost 50% of the total US toluene is dealkylated to benzene, the rest being used as a solvent or as a gasoline additive. Toluene, like benzene, is susceptible to substitution by many electrophilic reagents. It can be nitrated to trinitrotoluene, a well-known explosive, or to 2,4- and 2,6-dinitrotoluenes as important intermediates in polyurethane production.

Toluene can also be oxidized to benzoic acid. Fig. 1-5 shows the most important chemicals based on toluene.

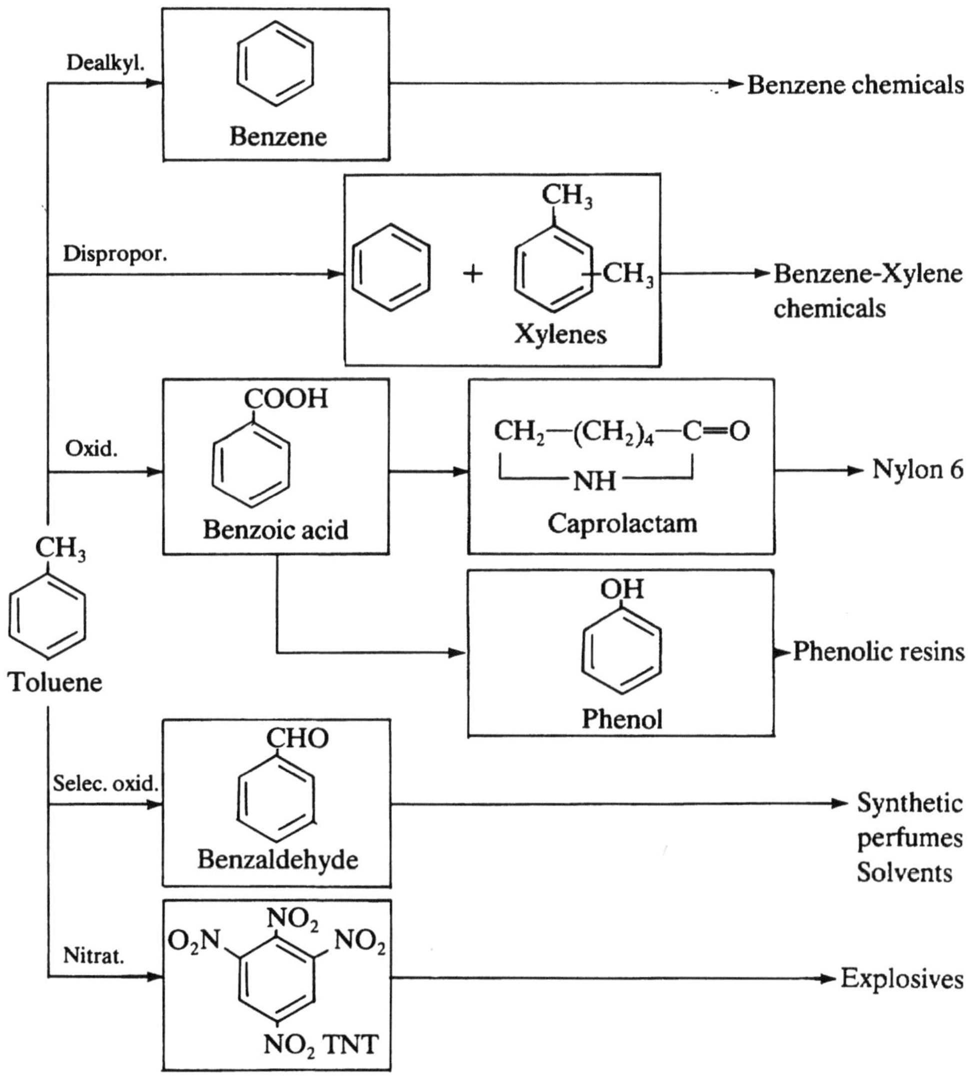

Fig. 1-5. Important intermediates and chemicals from toluene.

Xylene Chemicals

The mixture of xylenes from the reformates has three isomers: *ortho-*, *para-* and *meta*. The most valuable isomer is *p*-xylene. Separation of the three isomers is difficult, due to the close proximity of their boiling points. It is further complicated by the relatively low ratio of *p*-xylene (about 24 wt% of the total xylene). *p*-Xylene is the most important intermediate for producing terephthalic acid, which is used for the production of polyesters, the most important synthetic fiber raw materials.

Catalytic oxidation of *o*-xylene gives phthalic anhydride for the production of plasticizers. *m*-Xylene, although the major component of the xylene mixture from catalytic reforming, is the least utilized of the three isomers for chemical production. It is usually isomerized to the other two isomers. Fig. 1-6 shows the most important chemicals based on BTX.

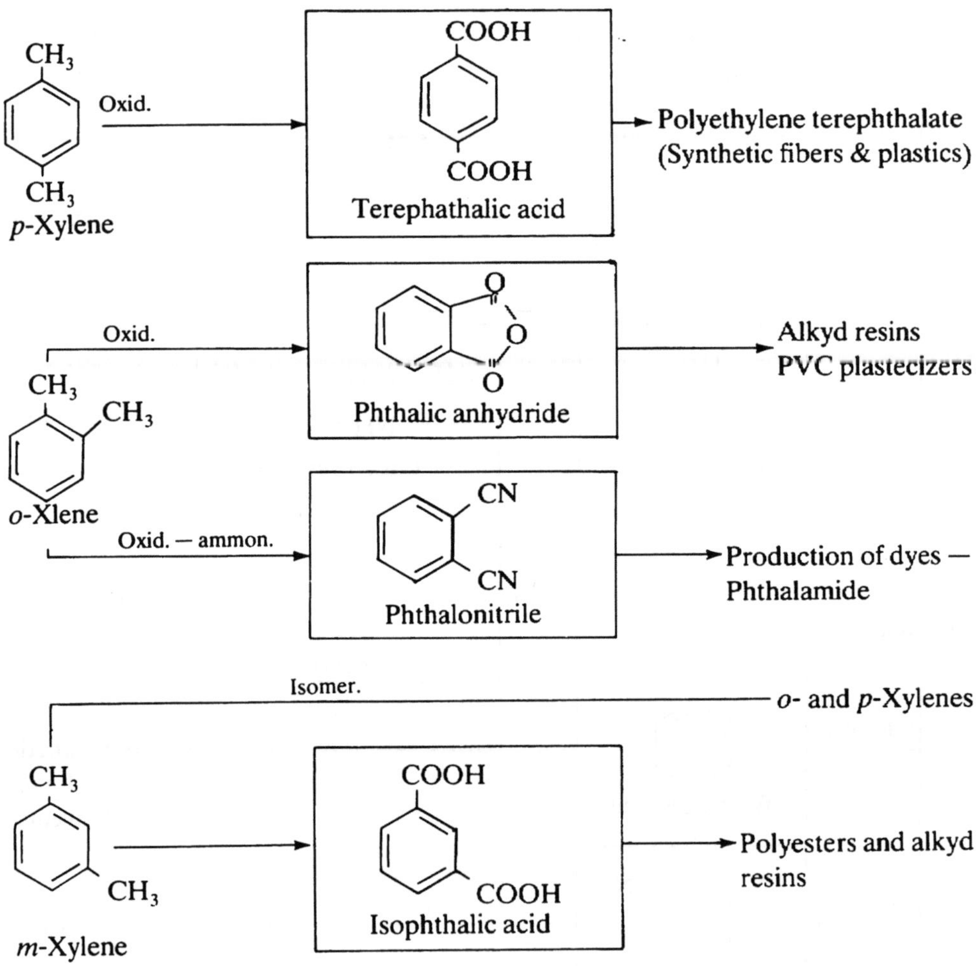

Fig. 1-6. Important chemicals from xylenes.

1.2.4. SYNTHESIS GAS

As mentioned above, synthesis gas can be produced by steam reforming of natural gas (methane).

$$CH_4 + H_2O \rightarrow CO + 3\ H_2$$

It can also be produced by steam reforming of other petroleum fractions, but the H/CO ratio is lower than that of natural gas. Gasification of coal and peat produces a mixture of carbon monoxide and hydrogen and the mixture is called medium BTU gas. This mixture, which may be used as an alternative energy source, could also be utilized as a chemical intermediate (synthesis gas).

The following is a summary of the reactions and the most important chemicals from synthesis gas.

A — Methanol Production

$$CO + 2\ H_2 \rightarrow CH_3OH$$

B — Hydrocarbons Production (Fischer Tropsch Synthesis)

$$n\ CO + 2n\ H_2 \rightarrow -(CH_2)_n- + n\ H_2O$$

$$n\ CO + 2n\ H_2 \rightarrow C_nH_{2n} + n\ H_2O$$

$$2n\ CO + n\ H_2 \rightarrow C_nH_{2n} + n\ CO_2$$

($n > 1$ for the last two reactions.)

C — Aldehydes and Alcohols (OXO Synthesis)

$$CH_2{=}CH_2 + CO + H_2 \rightarrow CH_3{-}CH_2CHO \xrightarrow{H_2} CH_3CH_2CH_2OH$$

D — Carboxylic Acids (Carbonylation of Alcohols)

$$ROH + CO \rightarrow RCOOH$$

$$CH_3{-}OH + CO \rightarrow CH_3{-}COOH$$

E — Vinyl Acetate (Carbonylation of esters)

$$2\ CH_3\overset{\overset{O}{\|}}{C}OCH_3 + 2\ CO + H_2 \rightarrow CH_3{-}CH(OCCH_3)_2 + CH_3COOH$$

$$CH_3{-}CH(OCCH_3)_2 \xrightarrow{\Delta} CH_2{=}CHOCCH_3 + CH_3COOH$$

F — Ethylene Glycol

$$2\ CO + 2\ H_2 \xrightarrow{\text{Rh catalyst}} H\overset{\overset{O}{\|}}{C}CH_2OH \xrightarrow{H_2} HOCH_2CH_2OH$$

G — Hydrogen for Ammonia Synthesis

$$CO + H_2 + H_2O \rightarrow 2\ H_2 + CO_2$$

Hydrogen freed from carbon dioxide by absorption in an alkaline solution such as diethanolamine is used for ammonia synthesis

$$3\ H_2 + N_2 \rightarrow 2\ NH_3$$

References

Hatch, L. and Matar, S.: *From Hydrocarbons to Petrochemicals.* Gulf Publishing Co., Houston, Texas, 1981.

Weissermel, K. and Arpe, H.: *Industrial Organic Chemistry.* Verlag Chemie, Weinheim, 1978.

Matar, S.: *Synfuels — Hydrocarbons of the Future.* PennWell Publishing Co. Tulsa, Oklahoma, 1982.

Brownstein, A.: *Trends in Petrochemical Technology.* Petroleum Publishing Co., Tulsa, Oklahoma, 1976.

Heaton, C. (Editor): *An Introduction to Industrial Chemistry.* Leonard Hill, Glasgow, 1984.

CHAPTER 2

Types of Catalysts used in Petrochemical Processes

2. Introduction

Catalysis is a phenomenon related to reaction rates, i.e. to reaction kinetics. Consider the simple combustion of natural gas:

$$CH_4 + 2\ O_2 \rightarrow CO_2 + 2\ H_2O$$

The reaction is highly exothermic; a fact that made possible the use of natural gas as a convenient source of energy. Thermodynamic principles suggest that a mixture of ethane and oxygen should react spontaneously to produce the lower-energy, more-stable CO_2 and H_2O. However, it is known that the combustion of ethane does not take place unless we provide the reaction mixture with some energy, by striking a match, for instance. Likewise, living organisms should burn spontaneously in air, were their existence to be governed by thermodynamic factors only. The fact is that another factor has to be taken into consideration and that is the reaction path. Although the CH_4—O_2 mixture is higher in energy than the product mixture (CO_2—H_2O), Fig. 2-1, still the reactants will not simply roll down to the more stable product position. The reactant mixture has to climb an energy hill (activation energy, which is always higher than the energy of the products or the reactants) before it can form products.

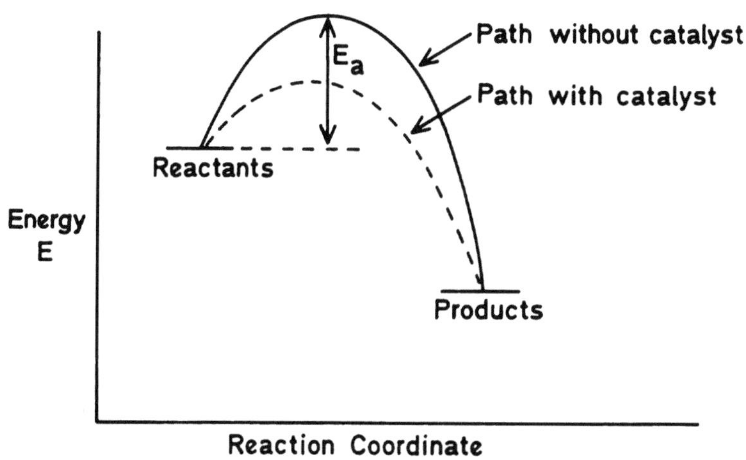

Fig. 2-1. Activation Energy E_a, case of an exothermic reaction.

The higher the energy hill that the reactants have to climb (the greater the activation energy E_a) the more difficult it is for the reaction to proceed (the slower the reaction). The function of a catalyst is to provide a path of lower activation energy (a tunnel in the energy hill), thus making it easier for reactants to proceed to products (the reaction is faster).

Catalysis is a very common phenomenon in chemical systems. Biological processes in living organisms take place under mild conditions by the aid of catalysts (generally called enzymes). Catalysis in the chemical industry causes the realization of otherwise very difficult reactions; and it makes the reaction conditions milder than if the catalyst were absent, thus resulting in great energy savings.

Although the catalyst is involved in one or more of the elementary reaction steps, it does not appear in the overall equation of the reaction. A basic requirement in catalytic reactions has been, therefore, that it should be possible to separate the reaction products from the catalyst. The longer the catalyst retains its original identity the longer it will remain active. If it enters into reactions leading to products from which the catalyst may not be regenerated, the catalyst may lose its activity. It is then said to be deactivated. A catalyst may react with an impurity in the reaction system to form a product that has no catalytic activity. It is then said to be poisoned. Catalyst poisoning and deactivation are among the major problems that the chemical industry tries to combat.

2.1. Catalyst Types

A catalyst may be indistinguishable from the reaction medium, forming one phase with other components of the medium. It is then called homogeneous. Homogeneous catalysts are usually used in liquid media in which they dissolve completely. If the catalyst exists in a phase different from that of the reaction medium, then it is termed 'heterogeneous'. Heterogeneous catalysts used in the petrochemical industry are generally solids while the reactants are either gases or liquids.

2.1.1. HETEROGENEOUS CATALYSTS

Heterogeneous catalysts are more common in the petrochemical industry than homogeneous catalysts, although their chemistry is less well-understood. It is no exaggeration to state that the level of our understanding of some homogeneously catalyzed reactions after five years of study is equivalent to the level of our understanding of heterogeneously catalyzed ones after 50 years of study. The reasons for this are simple. Homogeneous systems are well-behaved and can be investigated by simple physical methods, e.g. spectroscopically, that allow the detection of intermediates and the investigation of the reaction kinetics. The study of heterogeneous systems requires sophisticated instrumentation. The results are not always conclusive, because the reactions are difficult to reproduce.

Heterogeneous systems have been favored by industrialists because they pose simpler engineering problems. Continuous operations are possible here and separation of reaction products is simple. The gaseous reactants (for instance) are allowed to pass over a solid catalyst where the reaction takes place and products emerge as gases that may be separated and condensed, etc.

Heterogeneous catalysts still account for the production of the major part of petroleum and petrochemical products. They account for all the catalysts used in the refining processes and for about 80% of those used in the petrochemical industry. These include: ammonia synthesis, methanol production, acrylonitrile, styrene, vinyl chloride, vinyl acetate and butadiene manufacture.

2.1.2. HOMOGENEOUS CATALYSTS

The chief limitation for the potential use of heterogeneous catalysts is their lack of specificity. It is precisely in this area where homogeneous systems are superior to heterogeneous catalysts. Because of the small size of the catalytic entity, only one functional group of a poly-functional molecule is likely to be engaged at any one time. Moreover, the coordination requirements of the metal atom in a catalytic compound are likely to render any such interaction quite specific. On the surface of a heterogeneous catalyst, on the other hand, the simultaneous engagement of more than one function is frequently possible, resulting in nonselective behavior.

Homogeneous catalysts operate under much milder conditions than heterogeneous catalysts, thus reducing the energy cost. Since the more expensive Pt, Pd and Rh elements are among the most effective catalysts, more efficient use of the catalyst (by applying a homogeneous system) will result in more favorable process economics.

However, homogeneous catalysts suffer from the difficulty of their separation from the reaction medium.

Among the major homogeneously catalyzed processes that have been successfully commercialized in the petrochemical industry we may mention:

(a) The Wacker process for the oxidation of ethylene to acetaldehyde.
(b) The OXO processes for the manufacture of alcohols.
(c) The Monsanto process for the manufacture of acetic acid from methanol.

2.1.3. HETEROGENIZED HOMOGENEOUS CATALYSTS

The advantages of homogeneous and heterogeneous catalysis have recently been combined in systems that use heterogenized homogeneous catalysts. These systems have opened new areas for catalytic research. Features such as high selectivity, homogeneous active sites, and ease of chemical modification have been retained in these heterogenized homogeneous catalysts. The advantages of heterogeneous catalysis, such as ease in catalyst removal, recovery and reactivation, durability, and thermal or air stability, have also been retained.

Organic polymers such as polystyrene, polypropylene or polyvinylchloride, and inorganic polymers such as silica, alumina or zeolites are used as heterogenizing agents. The catalyst metal atom is anchored to these polymers through a group that is chemically bound to the polymer and containing a coordinating site e.g. —PPh_2 or —C_5H_4 (cyclopentadienyl) groups.

The catalytic activity of the metal moiety is invariably influenced by its attachment to the polymer chain. The catalytic activity of a polystyrene-bound (P)$RhCl(PPh_3)_2$ catalyst for hydrogenation, for example, exceeds that of the free $RhCl(PPh_3)_3$ primarily because steric crowding caused by the polymer chains maintains coordinative unsaturation of the metal atom during catalysis. Typical examples of heterogenized homogeneous catalysts (sometimes called polymer-bound or anchored catalysts) are shown in Table 2-1.

Regioselectivity in the hydroformylation reaction has been influenced by the use of a polymer-attached $RhH(CO)(PPh_3)_3$ catalyst, as shown in the example below:

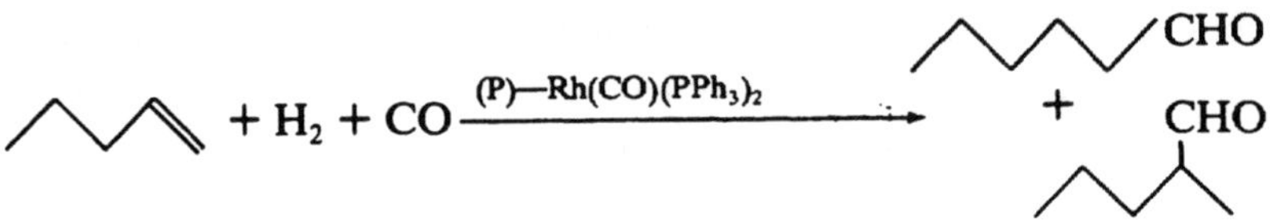

Ratio of normal to branched isomer = 16.1 (at 120°).

Table 2-1. Some examples of polymer-bound metal complexes.

Cross-Linked Polystyrene as the Polymer

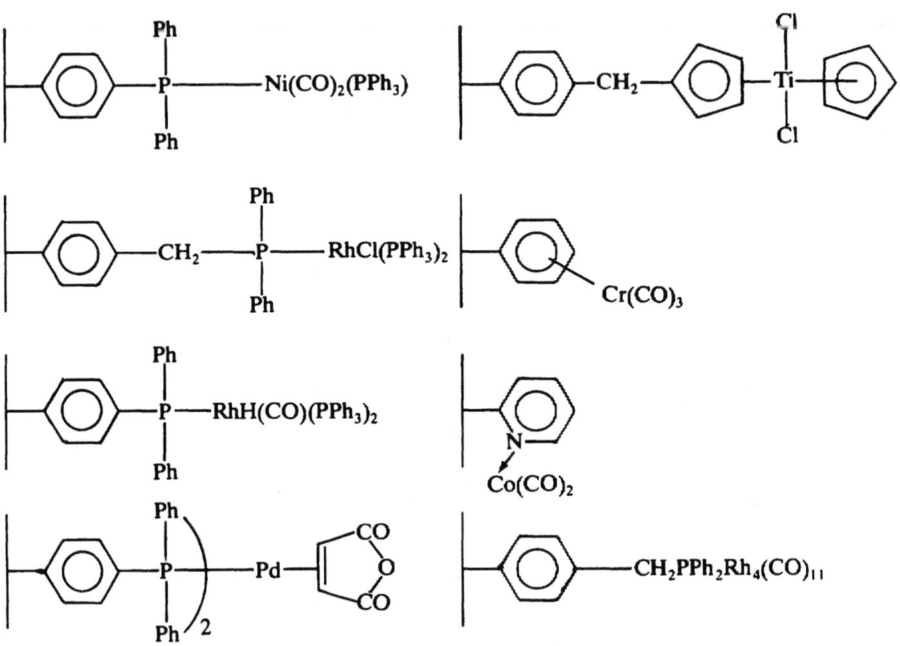

Table 2-1 (continued)

Silica or Alumina as the Polymer

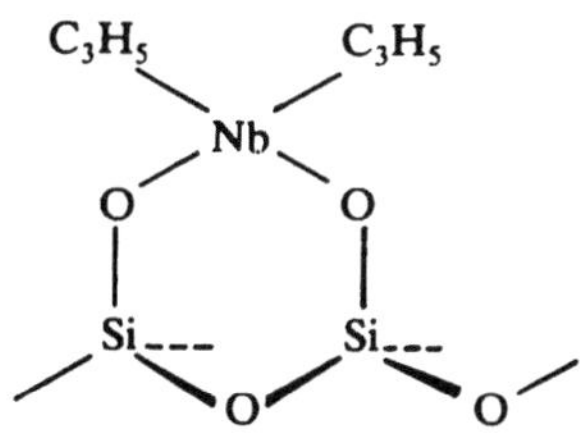

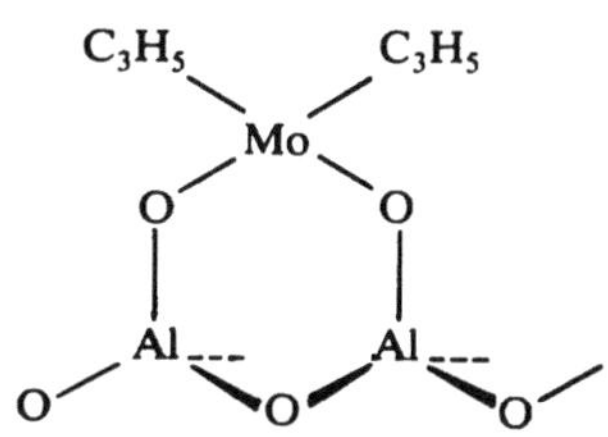

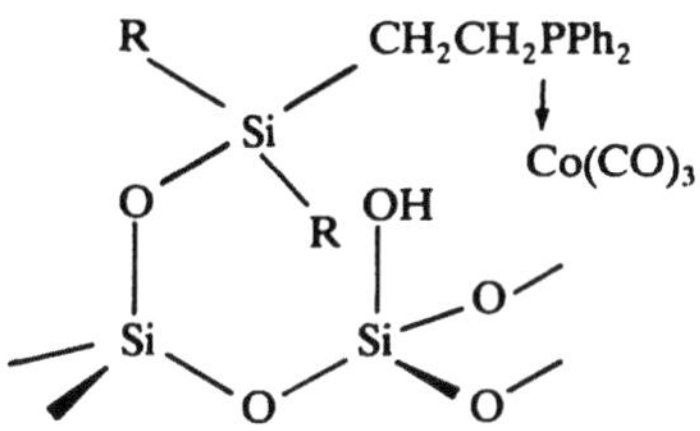

2.2. Basic Principles of Coordination Compounds

The catalytic reaction, whether homogeneous or heterogeneous, invariably involves, in one or more of its elementary steps, the formation of some coordination compound from the metal atom and the reactants. An understanding of the basic principles of coordination compounds is therefore a very important prerequisite for the understanding of the chemistry of the catalytic processes in the petrochemical industry. In order to master the principles of coordination chemistry one must be familiar with the development of the periodic table, electronic configurations of the elements, factors affecting the stability of various oxidation states, and theories of ionic and covalent bonding.

2.2.1. THEORIES OF BONDING IN COORDINATION COMPOUNDS

Coordination compounds have always been a challenge to the inorganic chemist. In the early days of chemistry they seemed unusual (hence the name 'complex' ions) and seemed to defy the usual rules of valence. Today they comprise a large body of current inorganic research. A survey of articles in recent issues of the journal *Inorganic Chemistry* indicates that perhaps 70% of the articles deal with coordination compounds. Although the usual bonding theories can be extended to accommodate these compounds, they still provide stimulating problems to be resolved. In synthetic work they continue to provide a challenge in the laboratory.

The rapidly developing field of bioinorganic chemistry focuses the presence of coordination compounds in living systems.

The modern study of coordination compounds started at the turn of the century with Alfred Werner (Nobel Prize, 1913). He introduced a set of postulates that explained the phenomenon of coordination numbers 4 and 6 and the geometry of coordination compounds.

Valence Bond Theory (VBT)

Linus Pauling (Nobel Prize, 1954) successfully applied the electron-pair bonding theory to coordination compounds. The theory consists of the following postulates:

(a) A metal-ligand bond is a result of the reaction of a Lewis base (ligand) with a Lewis acid (metal).
(b) The formation of the metal—ligand (M—L) bond therefore requires the availability of a lone pair of electrons on the ligand and an empty atomic orbital in the metal.
(c) The metal atom usually uses hybridized orbitals. Hybridization enhances the directional property of the orbital and results in greater extent of M—L orbitals overlap and consequently more stable bond.
(d) The geometry of the coordination compounds is generally octahedral (d^2sp^3 hybridization) for 6-coordinate complexes and tetrahedral (sp^3) or square-planar (dsp^2) for 4-coordinate species.
(e) The experimental differentiation between dsp^2 and sp^3 hybridizations in tetracoordinate d^8 complexes can be achieved by measuring the magnetic moment of the complex (dsp^2 compounds are diamagnetic while tetrahedral compounds are paramagnetic).

Although the VBT has many practical advantages it is far from being complete. It offers after-the-fact interpretation of experimental observations. It fails to explain the spectra of coordination compounds. It suffers from several other deficiencies.

Simple Electrostatic Theory

This theory describes the bonding between the metal atom (or ion) and the ligand in terms of pure electrostatic attraction between a positive charge centered at the nucleus of the metal atom and negative charges on the ligand atoms bonded to the metal. The negative charge could be due to an anionic ligand (e.g. halide) or the negative end of a polar ligand (e.g. H_2O or NH_3). However, this theory fails to account for the formation of highly stable metal carbonyls in which the metal atom is neutral and the ligand (CO) is almost non-polar. Moreover, the variation of the M—L bond strength (as measured by ΔH_f) for the same ligand over a series of M^{+n} along a transition-metal series shows that the correlation is good only for the ions with configurations d^0, d^5 and d^{10}. Other ions deviate from a straight-line relationship. This suggested that a modification of the electrostatic model is necessary.

Crystal Field Theory (CFT)

This is a modified electrostatic theory. It considers the metal and ligands as point charges. However, it also takes into consideration the electrostatic interaction (repulsion) between the electrons of the incoming ligands and those of the valence shell of the metal atom. Such a repulsion causes the electrons of the metal to seek atomic orbitals that are relatively out of the paths of the approaching ligands. The metal *d*-orbitals are therefore no longer degenerate. They split into two or more energy levels depending on their orientation with respect to the M—L bond direction. The splitting of *d*-orbitals in various geometries is shown in Table 2-2.

This model is a great improvement over all previous theories. It explains the magnetic properties of transition-metal complexes, the observed tendencies in stability and, qualitatively, the absorption spectra of transition metal complexes.

The theory suffers from some drawbacks. Calculations based on CFT do not yield results consistent with those obtained experimentally for electronic transition spectra of metal complexes. Moreover, the question why do some strong M—L bonds exist wherein the metal is in its (0) oxidation state and the ligand is non-polar, is still unsolved.

Table 2-2. The energy levels of *d* orbitals in crystal fields of different symmetries.[a]

C.N.	Structure	d_{z^2}	$d_{x^2-y^2}$	d_{xy}	d_{xz}	d_{yz}
2	Linear[b]	10.28	−6.28	−6.28	1.14	1.14
3	Trigonal[c]	−3.21	5.46	5.46	−3.86	−3.86
4	Tetrahedral	−2.67	−2.67	1.78	1.78	1.78
4	Square planar[c]	−4.28	12.28	2.28	−5.14	−5.14
5	Trigonal bipyramid[d]	7.07	−0.82	−0.82	−2.72	−2.72
5	Square pyramid[d]	0.86	9.14	−0.86	−4.57	−4.57
6	Octahedron	6.00	6.00	−4.00	−4.00	−4.00
6	Trigonal prism	0.96	−5.84	−5.84	5.36	5.36

[a] All energies are in multiples of certain energy unit (Dq), taking the energy of unsplit *d* orbitals in the absence of the crystal field as zero.
[b] Ligands lie along *z* axis.
[c] Ligands lie in *xy* plane.
[d] Pyramid base in *xy* plane.

Molecular Orbital Theory (MOT)

The existence of complexes of metal atoms in their zero or negative oxidation states with non-polar ligands, and the available evidence to the occurrence of electron sharing between M and L even in the most ionic of complexes (CoF_6^{3-}) suggest that a model based on covalent bonding is called for. Such a model should not suffer from the deficiencies of Pauling's model. The molecular orbital theory presents such a solution. It is not possible to apply the (MOT) in its exact form to complicated systems such as the coordination compounds. The theory is therefore simplified by considering only valence shell orbitals of the metal atom and the atomic or molecular orbitals accommodating the lone pair of electrons on the donor site of the ligand in addition to empty low-energy atomic or molecular orbitals on the ligand. The application of symmetry considerations will further simplify the problem. It is interesting to observe that the application of symmetry tools results in the splitting of the *d*-orbitals in a way qualitatively equivalent to the splitting resulting from the application of CFT.

The simplified MOT affords results consistent with experimental observations. Treatment of M—L bonds between low-valent metal atoms and non-polar ligands according to the MOT resulted in the introduction of the principles of back-donation of electrons from the metal to the ligand in molecular π-orbitals. Such interaction results in influencing the activity of the ligands and thus has far-reaching effects on the fate of the catalytic reaction.

Although the MOT is superior to other theories, the latter still have some applications in situations requiring simple treatment and known to constitute no exceptions to those theories.

2.2.2. STRUCTURE AND REACTIVITY OF COORDINATION COMPOUNDS

The coordination number (number of ligands directly bonded to the central metal atom), the geometry of the coordination compound and its reactivity are intimately related properties that determine the catalytic behavior of the coordination compound. A metal atom that assumes, under certain conditions, a coordination number which is different from the most stable one for that configuration, will have a tendency to change to the more stable coordination number once the environmental constraints are removed. Thus a tetracoordinate Pd(II) atom in PdL_2HX may be tempted to bond to a fifth ligand (an olefin in a catalytic reaction) to form $PdL_2HX(C_2H_4)$. However, such a pentacoordinate complex is unstable with respect to conversion to a tetracoordinate configuration. This reactivity drives the complex to undergo a hydride shift and form the tetracoordinate (albeit unstable) complex $PdL_2X(CH_2CH_3)$. The choice of the coordination number and the prefered geometry are determined by factors related to the metal atom and to the ligand as we shall see later.

Stereochemistry of Coordination Compounds

The maximum overlap requirement (VBT), minimum repulsion between adjacent ligands (Electrostatic Theory) and other energy considerations (CFT and MOT) result in predicting the known common shapes associated with the various coordination numbers: linear (2), triangular planar (3), tetrahedral or square planar

(4), trigonal pyramidal or square pyramidal (5), and octohedral (6). Whenever there is more than one possible geometry for the same coordination number (e.g., coordination number 4 and 5) with comparable stabilities, the more stable configuration is determined by the nature of the ligand and the metal atom.

Coordination numbers of particular interest in catalysis are 4, 5 and 6.

Coordination number 4: This is a very common number with two possible geometries: tetrahedral and square planar.

(i) *Tetrahedral Complexes*: These are favored by large steric requirements of the ligand and by metal atoms with configurations that do not involve appreciable crystal field stabilization energy (CFSE). They are therefore encountered with complexes of halides with small metal ions of (a) noble gas configuration e.g. Be^{++}, (b) pseudonoble gas configuration e.g. Zn^{++}, Ga^{+3} and (3) transition metal ions which do not strongly favor other structures by virtue of the CFSE e.g. Co^{++} (d^7).

(ii) *Square Planar Complexes*: These are less favored sterically than tetrahedral complexes i.e. there is less space available for the ligand around the metal atom. They are therefore prohibitively crowded by large ligands. Moreover, if the ligands are small enough to form square planar complexes, an octahedral complex with two additional σ-bonds can form with little or no additional steric repulsion. Best known examples are those of d^8 species such as Ni^{++}, Pd^{++}, Pt^{++}, Rh^{+}, Ir^{+} and Au^{+3}. The prerequisite for the stability of square planar complexes is the presence of non-bulky, strong-field ligands with π-bonding ability. If the metal atom is heavy such a requirement is easily satisfied. This is why all but few tetracoordinate complexes of Pt^{++}, Pd^{++}, Rh^{+} and Ir^{+} are square planar.

The cyanide ion is a strong-field ligand in virtue of its negative charge and its ability to π-bond and accept electrons back-donated from the metal. The halides, on the other hand, are weak-field ligands. We find therefore that $Ni(CN)_4^{=}$ is square planar while $NiCl_4^{=}$ is tetrahedral. Nickel(II) is expected to form tetrahedral and square planar complexes of comparable stability with ligands of intermediate field strength. Such ligands are useful for designing systems that are suitable for the investigation of structural (square planar $\rightleftharpoons$ tetrahedral) isomerism.

Coordination number 5: Nyholm and Tobe have shown that if electrostatic forces were the only forces operating in bonding, 5-coordinate compounds would always disproportionate into 4- and 6-coordinate species:

$$2\ MX_5^{n-5} \rightarrow MX_4^{n-4} + MX_6^{n-6}$$

However, covalent bonding is of great importance and calculations based on a pure electrostatic model should not be expected to hold strictly. This coordination number is very important in catalytic reactions as we shall see later. The geometry of 5-coordinate complexes may be trigonal bipyramidal, square pyramidal or distorted forms of them. Examples include $CuCL_5^{-3}$ (TBP) and $Ni(CN)_5^{-3}$ (SP) in $Cr(en)_3Ni(CN)_5 \cdot 1.5\ H_2O$.

Coordination number 6: This is the most common coordination number. The most common geometry for its complexes is the octahedral. Hexa-coordinate complexes may exhibit geometric and optical isomerisms. Almost every transition metal ion with oxidation state +3 or above forms octahedral hexacoordinate complexes. Such complexes are also common for many transition metals with oxidation states lower than +3.

Factors Affecting the Stability of Coordination Compounds

One important property of coordination compounds that are useful as catalysts is their ability to react with substrates to form new complexes of intermediate stability. Intermediate stability of a complex means that the complex is stable enough to be formed and reactive enough to undergo further reaction with another reactant subtrate in order to facilitate the formation of the desired product. An understanding of the factors affecting the stability of coordination compounds will enable an investigator to recognize potential catalysts and modify their chemical composition to improve their catalytic activity. Factors affecting the stability of coordination compounds may be related to the ligand or to the metal.

(a) *Factors Related to the Ligand.* Ligands have been arranged in a spectrochemical series according to their crystal field strength which parallels the stability of the M—L bond according to the crystal field theory. This order is

$$I^- < Br^- < SCN^- < Cl^- < NO_3^- < F^- < OH^- < OX^{-2} < H_2O < NCS^- < CH_3CN < NH_3 < \text{en} < \text{bipy} < \text{phen} < NO_2^- < \text{phosph} < CN^- < CO$$

Certain ligands are known to stabilize particular oxidation states. Thus the higher oxidation states of metals exist mainly as oxides or fluorides due to the high electronegativity of O and F. Lower oxidation states are generally stabilized by ligands that are capable of accepting electrons from the metal by back-donation. This is why metal carbonyls are encountered with metals in their zero ($Cr(CO)_6$, $Fe(CO)_5$, $Ni(CO)_4$) or negative oxidation states; e.g. $Fe(CO)_4^=$ and $Co(CO)_4^-$. The following order has been suggested for π-accepting capacity of ligands:

$$NO > CO = RNC = PF_3 > PCl_3 > PCl_2R > PClR_2 > P(OR)_3 > PR_3 = SR_2 > RCN > o\text{-phen}.$$

Bidentate and polydentate ligands can bond to metal ions through more than one donor atom, resulting in structures incorporating ring systems, thus imparting additional stability to the complex. Such extra stabilization is known as the chelate effect. Examples of chelates include $M(acac)_3$ and $M(en)_3^{+3}$ where acac = acetylacetonate ion and en = ethylenediamine. The chelate effect arises mainly as a result of entropy and statistical effects. Steric effects associated with the size of the ligand may have a significant effect on the stability of the complex.

(b) *Factors Related to the Metal.* The nature of the metal atom greatly affects the stability of its complexes. Its position in the periodic table will determine the extent of the crystal field stabilization that its complexes involve. A metal ion in the second series will stabilize a complex to an extent 50% more than would a metal atom in the same group and of the same oxidation state but from the first series. Moreover, metals rich in electrons (filled or approximately filled valence-shell *d*-orbitals) tend to stabilize complexes with π-accepting ligands through back-donation.

The availability of several empty valence shell *d*-orbitals (the early members of the transition series) enhances the formation of complexes with high coordination number.

2.2.3. SOME REACTION TYPES CHARACTERISTIC OF COORDINATION COMPOUNDS

A transition metal invariably exhibits more than one oxidation state and may form complexes of varying coordination numbers. The ability to change oxidation state and/or coordination number made it possible for coordination compounds to undergo some reactions that are typical of them. Such reactions generally constitute parts of the mechanisms of several chemical processes that are catalyzed by coordination compounds as we shall see shortly.

These reactions include: oxidative addition, reductive elimination and insertion reactions.

Oxidative Addition Reactions

Such reactions involve the reaction of a metal complex $(ML_n)^{m+}$ with a species XY to yield the complex $\{ML_n(X)(Y)\}^{m+}$ in which both the oxidation state and the coordination number have increased by 2.

$$ML_n^{m+} + XY \rightarrow ML_n(X)(Y)^{m+}$$

In order for the metal atom to undergo such a reaction it has to be unsaturated with respect to valence and/or coordination. A typical example is d^{10} and d^8 complexes (Pt(0) or Ir^+) which are oxidized to the d^8 and d^6 complexes (Pt^{+2} or Ir^{+3}).

Oxidative addition reactions constitute the principal method of forming M—C, M—H, M—X and M—M bonds. A typical example is Vaska's complexes:

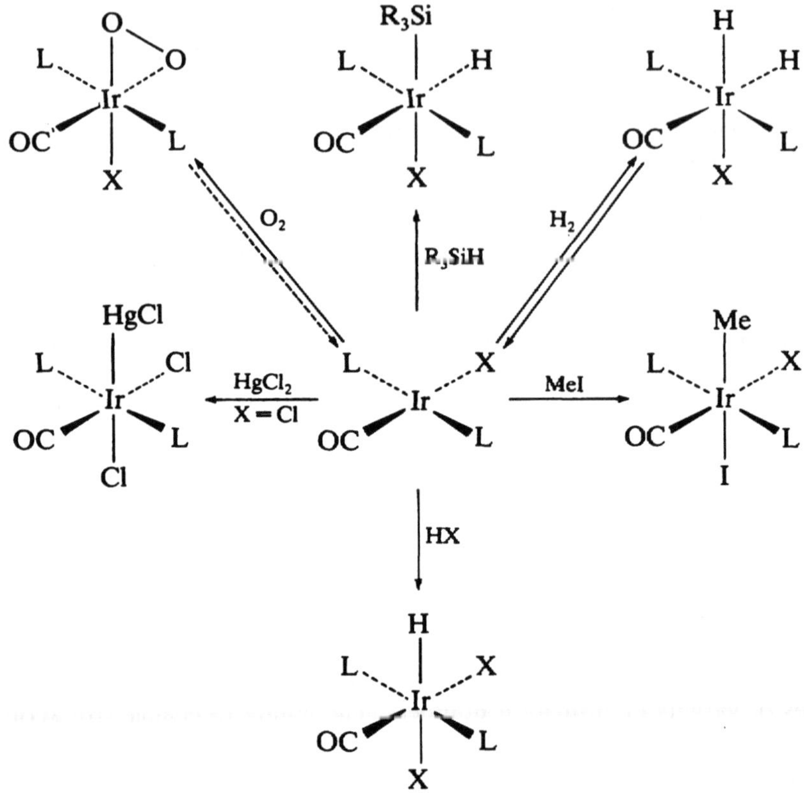

Another example is the reaction of $Pt(PPh_3)_2$ readily obtained by the dissociation of $Pt(PPh_3)_4$ in solution.

$$Pt(PPh_3)_4 \rightleftharpoons Pt(PPh_3)_2 + 2\ PPh_3$$

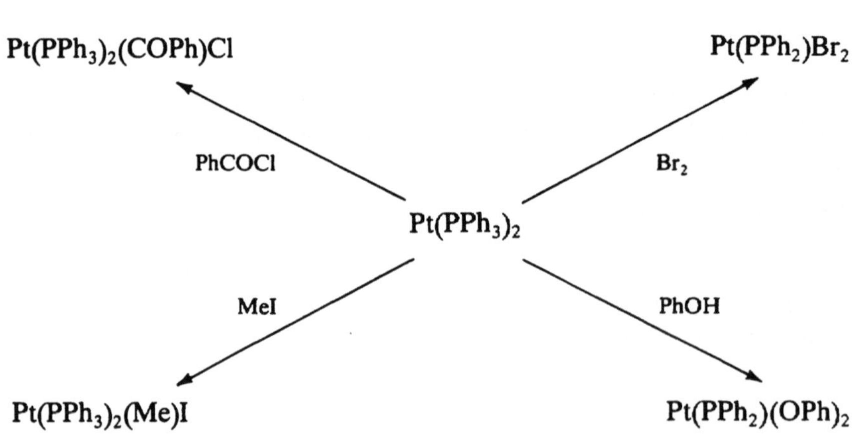

Several factors affect the reactivities of metal complexes towards most types of oxidative addition. The best studied cases involve d^8 and d^{10} complexes. Electron-releasing ligands such as phosphines increase, and π-acceptors such as CO or olefins decrease, the tendency to undergo oxidative addition. However, steric inhibition of oxidative addition is also quite important, especially with the bulky phosphines. Steric and electronic effects may be counterbalanced. Coordinatively unsaturated complexes are invariably more reactive than analogous saturated complexes. This factor is especially evident for d^{10} complexes of the type ML_n (M = Ni, Pd, Pt; and L = R_3P or RNC). In such cases $ML_2 > ML_3 > ML_4$.

The nature of the metal also influences its reactivity with respect to oxidative addition, which might have been expected to reflect the relative tendencies of the metals to become oxidized. Thus reactivity increases in the following order:

Fe(0) < Ru(0) < Os(0); Co(I) < Rh(I) < Ir(I);
Ni(II) < Pd(II) < Pt(II); Pt(II) < Ir(I) < Os(0).

Oxidative addition is one of the most important elementary reactions in inorganic chemistry. Many catalytically active transition metal complexes participate in oxidative addition. For example $RhCl(Ph_3)_3$ adds small molecules such as H_2, O_2, or CH_3I; and these small molecules are then activated for further reactions. Oxidative addition is a key step in homogeneous catalysis.

Reductive Elimination Reactions

Reductive elimination is the reverse of oxidative addition:

LMX—Y → LM + X—Y mononuclear (intramolecular)
LM—X + L—M—Y → L—M—M—L + X—Y dinuclear or intermolecular.

Reductive elimination is an obligatory step in virtually all reactions which are catalyzed by soluble transition metal complexes. It is therefore very important in homogeneous catalysis. It usually results in the formation of C—C, or C—H bonds.

Although reductive elimination has been studied to a much lesser extent than oxidative addition, the cases studied suggest that it is just as diverse and complicated as oxidative addition. The mechanisms of these reactions depend on the particular complex, the nature of the geometric disposition of the groups undergoing reductive elimination, the ancillary ligands, the coordination number, the oxidation state, and the formal charge on the complex.

It is difficult to distinguish an intra- from an intermolecular reductive elimination unless isotopic labels are used to deduce the presence or absence of cross-coupled products. Complexes which appear to be thermodynamically unstable with respect to reductive elimination, but are in fact kinetically inert, may exhibit a dinuclear path.

Mononuclear (intramolecular) reductive eliminations are favored by: (a) high formal charge on the metal, (b) the presence of bulky ligands, (c) *cis*-orientation of the two eliminating groups, (d) an electronically stable product complex, and (e) a coordinatively unsaturated intermediate.

Examples include:

$$RMe_2AuL \rightarrow Me{-}Me + R{-}AuL$$
$$RMe_2AuL \rightarrow RMe + Me{-}AuL$$

where $L = PPh_3$, $R = CH_3, CD_3$, Et, *n*-Pr.

On the other hand, dinuclear reductive elimination has been limited to eliminations involving a hydride. These reactions result in the formation of M—M, C—H or H—H bonds, e.g.

$$HMn(CO)_5 + MeAuPPh_3 \rightarrow Ph_3P{-}Au{-}Mn(CO)_5 + CH_4$$

Insertion Reactions

Insertion reactions are important as key steps in homogeneous catalysis as well as in stoichiometric organometallic syntheses. For example, all homogeneously catalyzed 'Oxo' reactions, olefin-oligomerizations, and olefin-hydrogenations, as well as heterogeneously catalyzed olefin-oligomerizations and hydrogenations and Fischer-Tropsch processes, involve insertion reactions.

Insertion reactions may involve the combination of a saturated ligand (X) with an unsaturated ligand (Y), forming a new ligand (Y—X):

$$X{-}M{-}Y \rightleftharpoons M{-}Y{-}X \rightarrow M{-}YX$$

This type of reaction is also called migratory insertion.

Examples of insertion reactions include migration of CO, olefins or acetylenes.

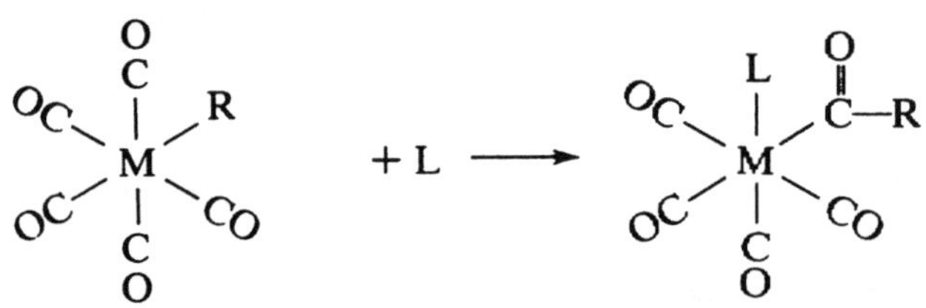

Metal alkyl → metal-acyl

The R group migrates to an adjacent carbonyl group, and, in a separate, subsequent step, the coordinatively unsaturated intermediate takes up an external ligand in the site vacated by the migrating alkyl group. This reaction is highly stereospecific. Alkyl groups having chirality at the M—C bond migrate with retention of configuration. The rate of these reactions and the position of equilibrium depend on the strengths of M—C and M—CO bonds. First-row $3d$ metal akyls are usually more reactive than $4d$ and $5d$ alkyls. Metals such as Pd(II) and Zr(IV) which form weak bonds to CO readily undergo the acyl-insertion reaction.

Another type of insertion involves the reaction of a metal complex with an external ligand or species whereby the new species is inserted in a M—X bond between the M and X e.g.

$$Pt(PPh_3)_2Cl_2 + SnCl_2 \rightarrow Pt(PPh_3)_2(SnCl_3)_2$$

Catalytic reactions may involve one or more of oxidative addition, reductive elimination and insertion reactions. However, all three reactions may be represented in one catalytic reaction. An example to such a case is the Monsanto Process for the manufacture of acetic acid from methanol using a rhodium compound as a homogeneous catalyst. The reaction can be represented by the following steps:

$$CH_3OH + HI \rightarrow CH_3I + H_2O$$

$$CH_3I + CO \xrightarrow{\{Rh(CO)_2I_2\}^-} CH_3C(O)I$$

$$CH_3COI + H_2O \rightarrow CH_3COOH + HI$$

The role of the catalyst is suggested to involve the following reactions:

$$\{Rh(CO)_2I_2\}^- + CH_3I \rightarrow \{CH_3RH(CO)_2I_3\}^- \qquad \text{oxidative addition}$$

$$\{CH_3Rh(CO)_2I_3\}^- \rightarrow \{CH_3CORh(CO)I_3\}^- \qquad \text{insertion of CO in Rh—CH}_3 \text{ bond}$$

$$\{CH_3CORh(CO)I_3\}^- + CO \rightarrow \{CH_3CORh(CO)_2I_3\}^-$$

$$\{CH_3CORh(CO)_2I_3\}^- \rightarrow \{Rh(CO)_2I_2\}^- + CH_3COI \qquad \text{reductive elimination}$$

2.2.4. REQUIREMENTS OF EFFECTIVE CATALYSIS

The discussion in this chapter could point to certain requirements that a metal atom should have in order to form compounds which would be potential catalysts. The requirements may be summarized in the following. It must be emphasized that such requirements are based on considerations related to homogeneous catalysis. It is safe to say, however, that they may be extended to heterogeneous catalysis in some respects.

(1) The metal atom should exhibit at least two oxidation states of comparable stability. Such a requirement will make it possible for its complexes to undergo oxidative addition and reductive elimination.

(2) The metal atom should be capable of attaining more than one coordination number, one of them being of lower stability. Of particular utility in catalysis is the ability of tetracoordinate metal atoms to form pentacoordinate complexes without change of oxidation state. The pentacoordinate complex, being of lower stability, will tend to form a tetracoordinate species and in doing so it will move the catalytic reaction one step forward.

(3) In order to effect selectivity in catalysis, a metal atom should have almost filled *d*-orbitals, but neither almost empty nor completely filled.
(4) It is highly favorable to use a less expensive catalyst whenever possible. It is unfortunate, however, that most of the effective catalysts are compounds of expensive metals.

The above requirements are typically satisfied by platinum and palladium in their +2 oxidation state. They have the d^8 configuration. They form stable tetraccordinate complexes in addition to moderately stable pentacoordinate ones. They exhibit +4 and 0 oxidation states in addition to the +2. It is not surprising therefore to see that many of the catalytic processes found to date use one form or the other of the complexes of these metals.

References

Huheey, J. E.: *Inorganic Chemistry: Principles of Structure and Reactivity*, 3rd Edition, Harper International SI Edition, 1983.
Collman, J. P. and Hegadus, L. S.: *Principles and Application of Organometallic Chemistry*, University Science Books, Mill Valley, Calif., 1980.
Satterfield, C. N.: *Heterogeneous Catalysis in Practice*, McGraw-Hill Book Co., 1980.

CHAPTER 3

Hydrogenation—Dehydrogenation Processes

3. Introduction

Hydrogenation is defined as the reaction between molecular hydrogen and an organic or an inorganic substrate. Hydrogenation reactions are exothermic but do not proceed at ordinary temperatures, except at negligible rates. The activation of hydrogen by the use of catalysts is among the oldest of chemical processes. Many metal oxides and metal complexes catalyze hydrogenation reactions. It may not be possible to list all hydrogenation reactions and the catalysts used, however, examples of industrial importance are discussed in this chapter. Although most commercial processes use hetrogeneous catalysts, homogeneous systems are gaining in importance for the production of high-value chemicals.

Dehydrogenation, on the other hand, is the loss of hydrogen from the substrate. Hydrogenation—dehydrogenation reactions are equilibrium systems and are generally activated by the same catalyst types.

$$H_2 + \text{substrate(1)} \underset{\text{dehydrog.}}{\overset{\text{hydrog.}}{\rightleftharpoons}} \text{substrate(2)}$$

Depending on the reaction conditions, hydrogenation may be favored over dehydrogenation or *vice versa*. A considerable degree of reaction control and selectivity, however, can be exercised by the choice of the catalysts to be used.

3.1. Hydrogenation

Hydrogenation is a hydrogen-consuming reaction, favored at high hydrogen partial pressure, while dehydrogenation is a hydrogen-releasing reaction favored at lower hydrogen partial pressure. In the hydroprocessing of petroleum fractions, hydrogen produced from dehydrogenation of naphtha (catalytic reforming) is used to hydrogenate undesirable compounds present in these fractions. For example, before naphtha is catalytically reformed, it is mildly hydrotreated to hydrodesulfurize and hydrodenitrogenate harmful sulfur and nitrogen compounds. These compounds are poisonous to the platinum catalysts used in catalytic reformers. Hydrodesulfurization and hydrodenitrogenation are examples of hydrogenolysis reactions where a hetero-carbon bond is cleaved.

Hydrogen is also used in other petroleum processes such as hydrocracking and hydrodealkylation. In some coal liquefaction processes, hydrogen is used to change the lower H/C ratio of coal (a solid fuel) to a higher ratio. In this way solid coal is hydrogenated to a liquid fuel known as synthetic crude. During the hydroliquefaction of coals, many complex reactions take place, such as cracking, hydrogenation of olefinic and aromatic compounds, and hydrogenolysis of nitrogen, sulfur and oxygen compounds. In addition to these applications, hydrogen is used in many synthetic processes which may require one or more hydrogenation steps.

3.1.1. HYDROGENATION CATALYSTS

Catalysts may generally be classified into heterogeneous and homogeneous. Typical examples of heterogeneous hydrogenation catalysts are:—

- Raney-metals (finely divided metals) e.g. Raney nickel.
- Noble metals on supports e.g. platinum on alumina, palladium on barium sulfate.
- Metal oxides e.g. Molybdenum oxide.
- Metal sulfides e.g. Tungsten sulfide.

Most, if not all hydrogenation—dehydrogenation catalysts are drawn from the transition metal series. These elements have a partially filled *d*-orbital. The catalytic activity of these elements has been partially attributed to the presence of vacancies in their electron *d* band which imparts an ionic-type bonding with electron transfer and conductivity. Another approach to explaining the catalytic activity is the power of these metals or their oxides to easily adsorb reactants and desorb products. A catalyst that strongly adsorbs a reactant molecule may not be a good catalyst, since it will not have the ability to yield it up easily to react with an adsorbed hydrogen atom. A catalyst that has a very weak adsorption power is also not a good catalyst since reactants will have surface concentrations that are too low to react at a reasonable rate. Only metals with a moderate adsorption power may prove to be good catalysts.

An important property of hydrogenation catalysts is their selectivity. By a proper choice of the catalyst, a reaction can be directed selectively toward a desired product. For example, copper chromite is known for carbon—oxygen bond hydrogenation while nickel is normally more used for carbon—carbon double bond saturation.

Homogeneous hydrogenation catalysts are usually in the form of metal complexes such as $ClRh(CO)(PPh_3)_3$ or metal carbonyls such as $Cr(CO)_3$ and $Co_2(CO)_8$. The catalytic activities of these complexes is better defined than those of the heterogeneous catalysts. Bonding can take place between a reactant's ligand-electrons and the metal ion which has a vacant or partially filled *d* orbital. An example of a homogeneous hydrogenation catalyst is hydrocobalt tetracarbonyl $[HCo(CO)_4]$. This compound can be prepared by reacting hydrogen with dicobalt octacarbonyl.

$$Co_2(CO)_8 + H_2 \rightleftharpoons 2\ HCo(CO)_4$$

Hydrocobalt tetracarbonyl acts as an acid in a polar solvent by giving up a proton, and as a hydride donor in a non-polar solvent. This complex is the catalyst used in the hydroformylation and carbonylation reactions.

One of the most elegant applications of homogeneous hydrogenation, is the synthesis of optically active compounds from nonchiral materials. The first commercial application of this approach is the production of a drug (L-Dopa) by Monsanto for treatment of Parkison's disease. Starting with a prechiral substituted cinnamic acid, a rhodium catalyst with an optically active phosphine ligand produces L-Dopa (a chiral product).

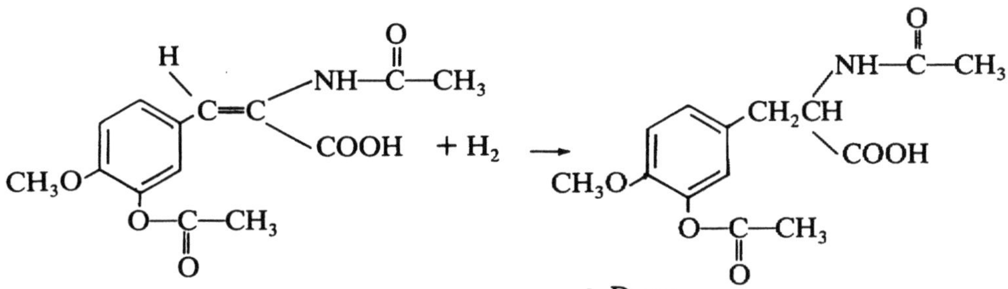

L-Dopa

The catalyst could be prepared by reacting a diene-rhodium ligand with trialkyl-phosphine and methyl alcohol.

$$HRh(diene)_2 + 2\,PR_3 + n\,CH_3OH \rightarrow [RhH_2(PR_3)_2(CH_3OH)_n]^+$$

3.1.2. HYDROGENATION REACTIONS AND PROCESSES

Hydrogenation reactions are those that consume hydrogen. Hydrogen is used to saturate either a multiple bond or a cleaving molecule (or an atom) which results from the breaking of such a bond. In the latter case the term hydrogenolysis is used.

Hydrogenation reactions are exothermic and are generally favored at lower temperatures. Depending on the activation energy of the compound to be hydrogenated, a reasonable reaction rate could be achieved at moderate temperatures by selecting the right catalyst. Higher operating pressures favor the forward reaction since the number of product moles is less than the number of reactant moles. A higher hydrogen partial pressure shifts the equilibrium toward the product side.

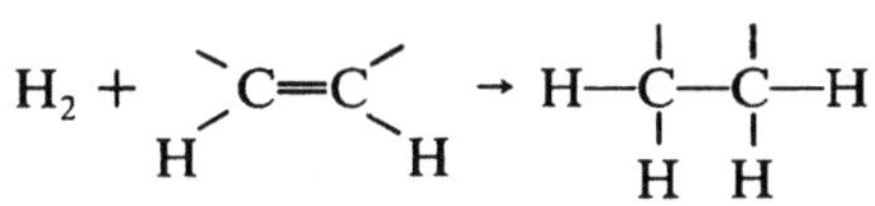

Operating at higher pressures, however, has little or no effect on hydrogenolysis reactions. For example, when hydrodesulfurizing a disulfide, the number of product moles equals the number of reactant moles, and increasing or decreasing the pressure does not affect the equilibrium.

$$R{-}S{-}S{-}R + 3\,H_2 \rightarrow 2\,RH + 2\,H_2S$$

Increasing the hydrogen partial pressure increases the hydrodesulfurization rate and shifts the equilibrium toward the desulfurized species.

Hydrogenation and hydrogenolysis reactions are extremely important when dealing with petroleum, petrochemical and coal processing. Reactions of industrial importance are discussed in this section. The objective and the usefulness of a hydrogenation reaction — which may represent only one step in the production sequence of a certain compound, as will be clearly stated in the course of the discussion.

3.1.2.1. *Hydrogenation of Alkenes and Alkynes (Production of 1,4-Butanediol)*

The hydrogenation or reduction of alkenes and alkynes is a simple addition reaction. The carbon—carbon bonds are broken and carbon—hydrogen bonds are formed, thus changing an active, unsaturated compound to a less active, saturated one. This reaction is not generally used to synthesize alkanes, but is more often utilized to saturate some olefinic compounds present in some petroleum fractions during the hydroprocessing of these fractions. However, with the choice of a selective catalyst, hydrogenation of alkynes may be stopped at the alkene stage. Hydrogenation of alkenes and alkynes may be utilized in the synthesis of some chemicals, as in the case of 1,4-butanediol. The hydrogenation reaction represents only one step in the production sequence of the diol.

1,4-Butanediol is an important chemical, mainly used for the production of tetrahydrofuran. It is also used in acetylenic chemicals and polyurethanes.

Acetoxylation of butadiene produces 1,4-diacetoxy-2-butene. This compound is then hydrogenated using a Ni—Zn catalyst on diatomaceous earth at about 80°C and 60 bar. The product, 1,4-diacetoxybutane, is hydrolyzed to the diol.

$$CH_3{-}C(=O){-}OCH_2{-}CH{=}CH{-}CH_2{-}O{-}C(=O){-}CH_3 + H_2$$

$$\rightarrow CH_3{-}C(=O){-}O{-}(CH_2)_4{-}O{-}C(=O){-}CH_3$$

$$CH_3{-}C(=O){-}O(CH_2)_4{-}O{-}C(=O){-}CH_3 + 2\,H_2O$$

$$\rightarrow HO{-}(CH_2)_4{-}OH + 2\,CH_3{-}C(=O){-}OH$$

Reduction of the triple bond is also used in synthesizing 1,4-butanediol via an alternative route. 2-Butyne-1,4-diol is prepared by reacting acetylene and formaldehyde. This is then reduced to the saturated diol using a Raney nickel catalyst at about 100°C and 300 bar.

$$HOCH_2{-}C{\equiv}C{-}CH_2OH + 2H_2 \rightarrow HO(CH_2)_4OH \qquad \Delta H = -251\ \text{kJ/mol}$$

3.1.2.2. *Hydrogenation of Aromatic Compounds*

Like alkenes and alkynes, hydrogenation of aromatic hydrocarbons involves breaking of the C—C π-bonds, and formation of C—H σ-bonds. However, more severe reaction conditions are needed for this hydrogenation due to the resonance energy associated with the aromatic nucleus. Examples of this type are the hydrogenation of:— benzene, benzoic acid and phenol for the production of ε-caprolactam and adipic acid.

a. *Production of ε-caprolactam*

ε-Caprolactam is the monomer used for producing nylon 6. Many routes are utilized for the production of the lactam, but they all start by hydrogenating the benzene ring to cyclohexane or substituted cyclohexane. Benzene, phenol or benzoic acid is hydrogenated to the corresponding saturated cyclic compound. These compounds are precursors of ε-caprolactam.

Liquid-phase hydrogenation of benzene at about 200°C and 30—50 bar produces cyclohexane.

Nickel, platinum or palladium on alumina are commonly used catalysts, but are all very sensitive to traces of sulfur compounds. Benzene used for the hydrogenation should not contain more than 1 ppm sulfur.

Gas-phase hydrogenation at about 500°C and 50 bar is currently practiced. To reduce the isomerization of cyclohexane to methylcyclopentane at this temperature, short residence times are required. An adequate heat removal system is needed since the hydrogenation reaction is highly exothermic ($\Delta H = -214$ kJ/mol).

Hydrogenation of phenol to cyclohexanol using a nickel catalyst takes place at about 150°C and 15 bar.

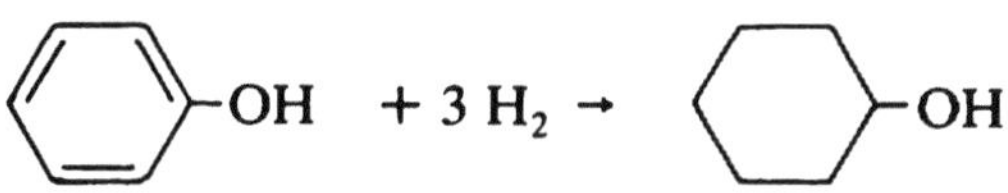

Cyclohexanol is not directly used for the production of ε-caprolactam but is further dehydrogenated in a second step at 400°C to cyclohexanone using a zinc or a copper catalyst.

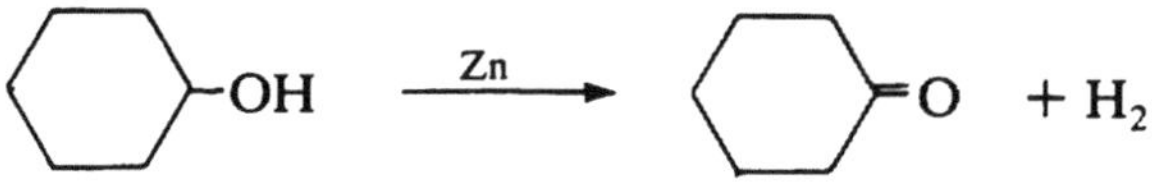

The net equation for the two reaction steps is:—

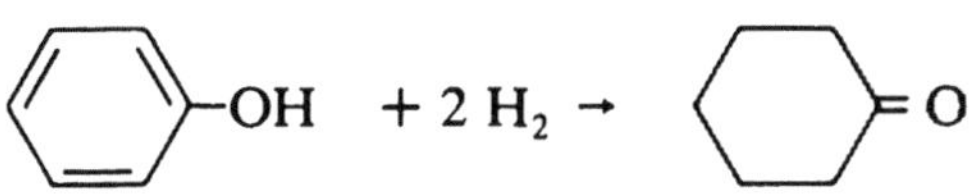

A highly selective, one-step, phenol-to-cyclohexanone process has been developed. The catalyst is palladium with some oxide such as CaO. The reaction is carried out at nearly atmospheric pressure at about 150°C.

Hydrogenation of benzoic acid to cyclohexane carboxylic acid is similar to phenol hydrogenation (approx. 150°C and 15 bar). The catalyst is palladium over charcoal.

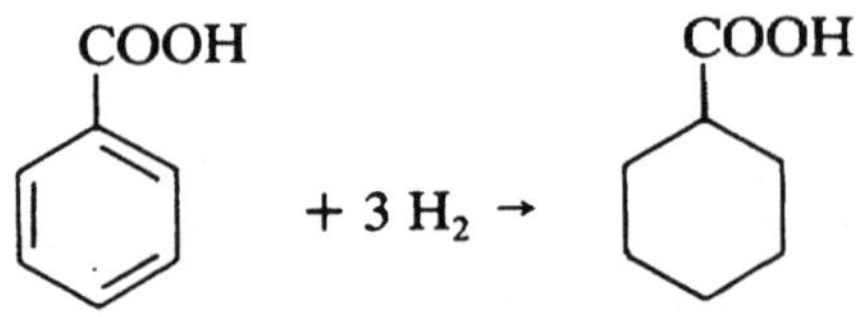

Any of the three hydrogenated compounds, cyclohexane, cyclohexanone and cyclohexane carboxylic acid, can be utilized for the production of ε-caprolactam. Using the appropriate reagent, cyclohexane or cyclohexanone can form the oxime.

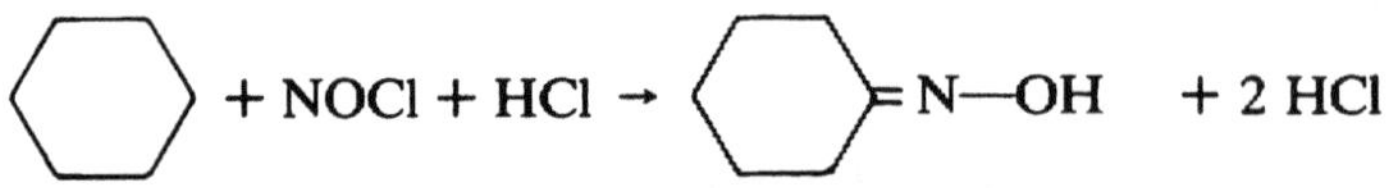

Treatment of the oxime with sulfuric acid produces ε-caprolactam.

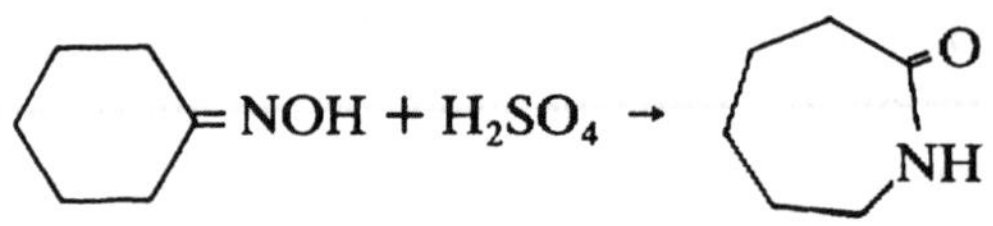

ε-Caprolactam can also be produced in one step using cyclohexane carboxylic acid by reacting it with nitrosyl hydrogen sulfate.

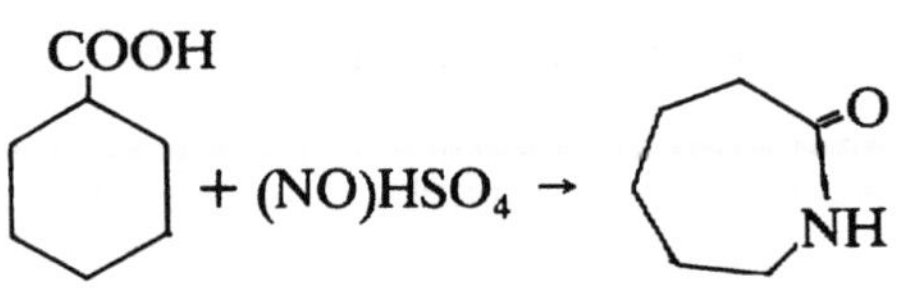

The older method for producing the oxime from cyclohexane, however, is by catalytically oxidizing it to a cyclohexanone—cyclohexanol mixture (KA oil). This mixture may be further reacted with nitrosyl chloride to give the oxime. The oxime could be used to produce ε-caprolactam as mentioned before.

b. *Production of adipic acid and hexamethylenediamine*

Adipic acid is an important monomer for the production of nylon 6/6. It is produced by the catalytic air oxidation of cyclohexane (hydrogenated benzene) in acetic acid.

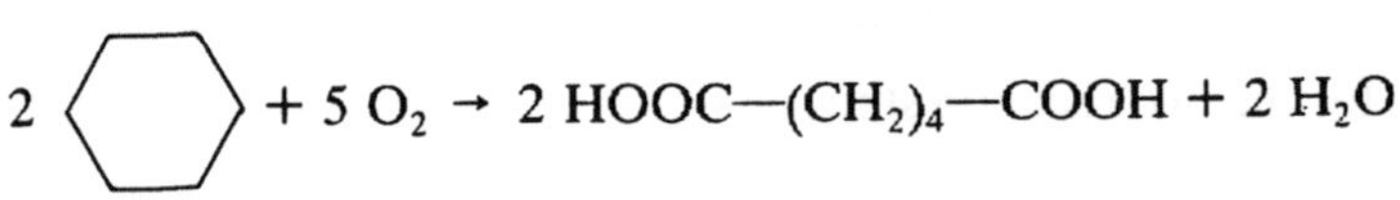

Adipic acid is also a precursor of hexamethylenediamine, the second monomer for nylon 6/6. In one process the acid is hydrogenated to 1,6-hexanediol which is further reacted with ammonia to give hexamethylenediamine.

$$HOOC(CH_2)_4COOH + 4\ H_2 \rightarrow HO{-}(CH_2)_6{-}OH + 2\ H_2O$$

$$HOH_2C(CH_2)_4CH_2OH + 2\ NH_3 \rightarrow H_2N(CH_2)_6NH_2 + 2\ H_2O$$

3.1.2.3. *Hydrogenation of Compounds with Heteroatoms*

In principle, all compounds having heteroatoms could be hydrogenated. In industrial applications, this reaction is important in hydroprocessing petroleum fractions and coal liquids that contain a high percentage of sulfur and nitrogen compounds. It is also useful in specific synthetic applications as in the reduction of a nitro to an amino group. Examples of industrial importance are given below.

A. *Hydrogenation of the nitrile group (production of hexamethylenediamine)*

Hexamethylenediamine, the other half of nylon 6/6, can also be produced by the catalytic hydrogenation of adiponitrile using a Cu/Co catalyst system at about 120°C and 600 bar. Nickel and nickel modified with iron or chromium are also useful catalysts for the reaction. An iron-based catalyst allows this reaction to take place at a lower pressure of about 300 bar at 150°C.

$$NC{-}(CH_2)_4{-}CN + 4\ H_2 \rightarrow H_2N{-}(CH_2)_6{-}NH_2 \qquad \Delta H = -314\ \text{kJ/mol}$$

Ammonia should be removed from the reaction medium since it suppresses the formation of the diamine. The formation of ammonia is favored under low hydrogen partial pressure from a competing reaction to that of diamine formation.

$$R{-}CN + H_2 \rightarrow R{-}CH{=}NH$$

$$R'CH_2NH_2 + R{-}CH{=}NH \rightarrow R{-}CH{=}N{-}CH_2{-}R' + NH_3$$

$$RCH{=}N{-}CH_2R' + H_2 \rightarrow RCH_2NHCH_2R'$$

A secondary amine

$$R = NC{-}(CH_2)_4{-}$$
$$R' = H_2N{-}(CH_2)_5{-}$$

Hexamethylenediamine may also be produced via a hydrodimerization reaction of acrylonitrile

$$H_2C{=}CH{-}CN + 2e \rightarrow H_2^{-}C{-}C^{-}HCN$$

$$H_2^{-}C{-}C^{-}H{-}CN + H_2C{=}CH{-}CN \rightarrow NC{-}{}^{-}CH{-}CH_2{-}CH_2{-}{}^{-}CHCN$$

$$NC{-}{}^{-}CH{-}CH_2{-}CH_2{-}{}^{-}CH{-}CN + 2\ H_2O \rightarrow NC{-}(CH_2)_4{-}CN + 2\ OH^-$$

Adiponitrile formed from the dimerization reaction can be further hydrogenated to hexamethylenediamine as mentioned above.

A competing reaction with dimerization is the early protonation of the anion radical intermediate to give propionitrile.

$$H_2^{-}C{-}C^{-}H{-}CN + 2\ H_2O \rightarrow CH_3CH_2CN + 2\ OH^-$$

The addition of a quaternary ammonium salt to the reaction medium increases the solubility of acrylonitrile in the aqueous medium. It also produces an aprotic layer adjacent to the electrode which inhibits the formation of propionitrile.

B. *Hydrogenation of the nitro group*

Aniline, is an important organic base used in many industries and particularly in the synthesis of dyes. It is produced by the liquid phase catalytic hydrogenation of nitrobenzene.

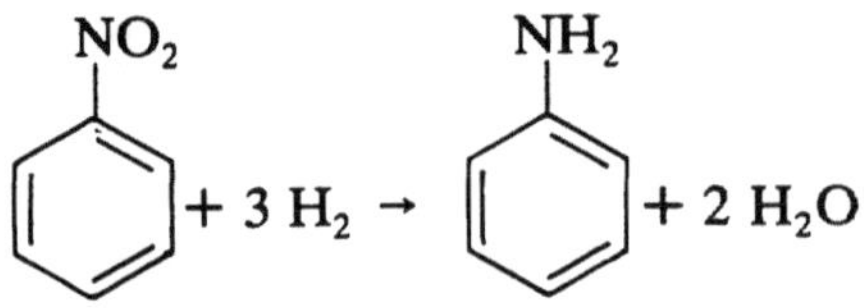

$\Delta H = 493$ kJ/mol

Selective hydrogenation of the nitro group without affecting the benzene ring could be done using a nickel sulfide catalyst activated with copper.

A gas-phase hydrogenation of nitrobenzone using a fluid-bed system under nearly atmospheric conditions is also practiced. The reaction takes place by the use of Cu/SiO catalyst in presence of excess hydrogen at about 280°C.

C. *Hydrogenation of the carbonyl group (production of alcohols)*

Aldehydes, ketones and carboxylic acids can be hydrogenated to the corresponding alcohols. Aldehydes and carboxylic acids produce primary alcohols and ketones produce secondary alcohols.

$$R-\overset{\overset{\displaystyle O}{\|}}{C}-H + H_2 \rightarrow R-CH_2OH$$

$$R-\overset{\overset{\displaystyle O}{\|}}{C}-R + H_2 \rightarrow R-\overset{\overset{\displaystyle OH}{|}}{C}H-R$$

Aldehydes produced by hydroformylation of olefins (Oxo aldehydes) are usually hydrogenated to the corresponding alcohols. The use of these alcohols depends on their chain length. The C-4—C-6 alcohols or their esters are used as solvents. Esters of C-8—C-13 alcohols are plasticizers and C-12—C-18 alcohols are used for the production of surface active agents. Nickel or copper catalysts are used for hydrogenating Oxo aldehydes. Either liquid or gas-phase processes are used. A pressure of approximately 80 bar and a temperature of about 120°C are used for the liquid phase process while a much lower pressure (about 2 bar) is used for the gas-phase reaction.

Both *n*-butyraldehyde and isobutyraldehyde are important Oxo aldehydes. Either cobalt or rhodium catalysts are used for their production from propylene. However, rhodium catalysts are preferred since they allow operation at much lower temperatures and pressures than when cobalt is used. Hydrogenation of *n*-butyraldehyde or isobutyraldehyde produces the corresponding alcohol.

The hydrogenation of adipic acid to 1,6-hexanediol has been mentioned before.

3.1.2.4. *Hydroprocessing Petroleum Products and Coal Liquids*

Petroleum fractions may contain large quantities of sulfur and nitrogen compounds. Using these products without prior treatment not only presents corrosion problems but also pollutes the environment. In addition, sulfur and nitrogen compounds poison many catalysts used in petroleum processing. Hydrotreatment of these fractions eliminates or reduces the sulfur and nitrogen compounds to an acceptable level. Hydrotreatment may also be necessary to saturate olefinic compounds present in some cracked fractions. Olefinic compounds are known to be precursors of coke formation, which is responsible for deactivating valuable catalysts. For example, cracked naphtha should be hydrotreated before it is catalytically reformed.

Coal liquids normally contain higher sulfur and nitrogen compounds than most petroleum fractions. More severe conditions are usually required for their treatment. This may also be necessary to partially hydrocrack the complex structure of the compounds normally found in coal liquids.

Although many reactions take place during hydroprocessing petroleum fractions and coal liquids, hydrodesulfurization and hydrodenitrogenation reactions are probably the most important ones. The process used for this purpose is usually called hydrodesulfurization.

A. *Hydrodesulfurization*

Hydrodesulfurization or hydrogenolysis of sulfur compounds produces a hydrocarbon and hydrogen sulfide. The most important classes of organic sulfur compounds found in petroleum fractions and coal liquids are mercaptans, sulfides, disulfides, thiophenes, benzothiophenes and dibenzothiophenes. The following are examples illustrating some of the reactions of these compounds during hydrodesulfurization.

Sulfur Class	*Reaction*
Mercaptans	$R{-}SH + H_2 \rightarrow RH + H_2S$
Sulfides	$R{-}S{-}R + 2\ H_2 \rightarrow 2\ RH + H_2S$
Disulfides	$R{-}S{-}S{-}R + 3\ H_2 \rightarrow 2\ RH + 2\ H_2S$
Thiophene	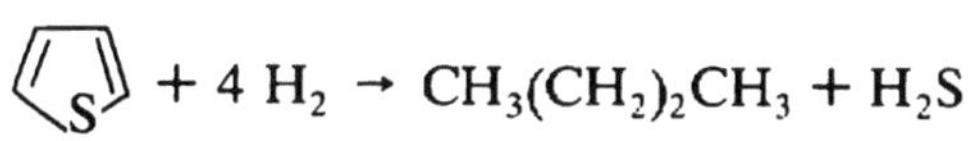
Benzothiophene	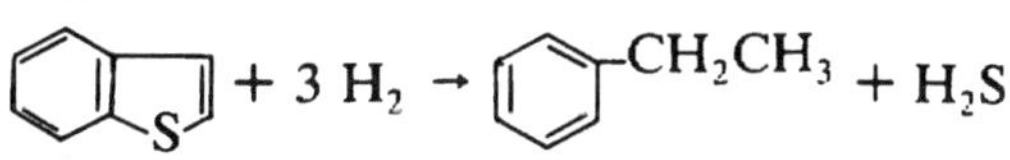
Dibenzothiophene	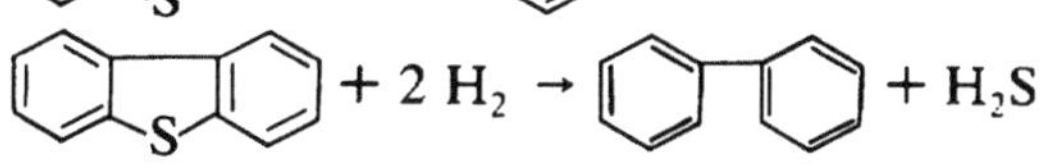

Comparing the hydrodesulfurization of different sulfur compounds, it has been generally found that sulfides and mercaptans are hydrodesulfurized more readily than thiophenes and benzothiophenes. This is expected, however, due to the aromatic character of the latter compounds.

When hydrodesulfurizing thiophene, the question arises whether hydrogenation takes place first before hydrogenolysis or *vice versa.* It has been proposed that butadiane (but not tetrahydrothiophene) is an intermediate in the hydrodesulfurization of thiophene. When tetrahydrothiophene was hydrodesulfurized, a different product composition than that of thiophene was obtained. This indicated that the C—S bond dissociation takes place before the hydrogenation step. In other words, tetrahydrothiophene may not be an intermediate in the hydrodesulfurization of thiophene while butadiene could be the intermediate.

Regarding benzothiophene, however, the rate of hydrogenation was found to be faster than the rate of hydrogenolysis. This led to the conclusion that dihydrobenzothiophene could be an intermediate in the hydrodesulfurization reaction.

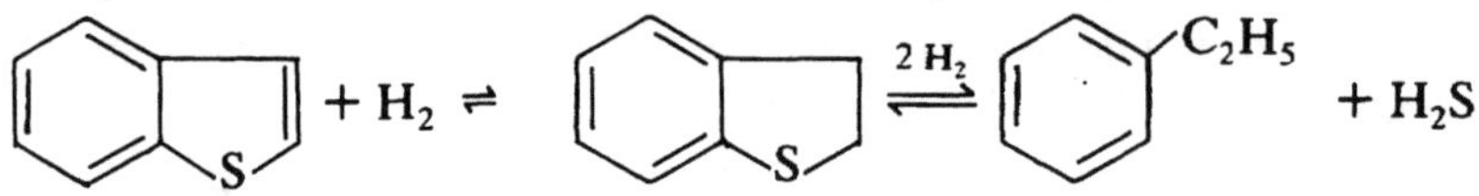

Hydrodesulfurization catalysts. Hydrodesulfurization requires catalysts which are sulfur resistant. The cobalt—molybdenum system was found to be effective in the desulfurization of petroleum products. Typical hydrodesulfurization catalysts are composed of about 4% CoO and 12% MoO_3 on γAl_2O_3 support. Before use, a catalyst is reduced and sulfided during the initial stages of operation using a H_2/H_2S mixture. Oxides are partially or completely changed to the sulfides. Some commerical catalysts are sold in a pre-sulfided form. In the sulfided form it is thought that some sulfur anions occupy oxygen anion positions in the outer catalyst layer. When sulfided, molybdenum oxide becomes reduced to +4 or +3 oxidation state. Other catalyst systems are also used in hydrodesulfurization processes such as NiO/WO_3. Nickel oxide may replace cobalt oxide and a NiO/MoO_3 system is a customarily used catalyst.

It has been found that neither cobalt nor nickel alone can activate hydrodesulfurization reactions to an appreciable extent. However, molybdenum or tungsten alone can support hydrodesulfurization reactions. Adding either nickel or cobalt to molybdenum or tungsten appreciably promotes their hydrodesulfurization activities.

A possible scheme for desulfurizing thiophene using the Co/Mo system where cobalt acts a promotor is illustrated in the following reaction sequence.

$$2\ Co^{2+} + 2\ H_2 \rightarrow 2\ Co^0 + 4\ H^+$$

$$2\ H^+ + 2\ S^{2-} \rightarrow 2\ SH^-$$

$$4\ Mo^{3+} \rightarrow 4\ Mo^{4+} + 4e^-$$

$$+ 4e^- + 2\ SH^- \rightarrow CH_2{=}CH{-}CH{=}CH_2 + 3\ S^{2-}$$

$$S^{2-} + 2\ H^+ \rightarrow H_2S$$

Net equation

$$2\,Co^{2+} + 4\,Mo^{3+} + 2\,H_2 +$$ 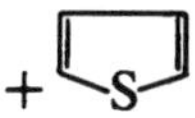 $$\rightarrow 2\,Co^{0} + 4\,Mo^{4+} + C_4H_6 + H_2S$$

Electron transfer for the catalyst and the promoter could be represented by a redox equation.

$$Co^{2+} + 2\,Mo^{3+} \rightarrow Co^{0} + 2\,Mo^{4+}$$

Sulfur compounds to be desulfurized (e.g. thiophene) are adsorbed on certain sites (anion vacancies) on the catalyst surface. By a reductive adsorption, a cation such as Co^0 may accept electrons from hydrogen and is changed to Co^0. Released protons react with surface sulfur anions forming SH^- anions. Meanwhile, Mo^{3+} becomes oxidized to Mo^{4+} by releasing electrons. Adsorbed sulfur compounds are then hydrodesulfurized by reacting with SH^- anions and the electrons released from the previous step. Hydrogen sulfide is formed by the reaction of a released sulfur anion with two protons from the reductive adsorption step.

B. *Hydrodenitrogenation*

Hydrodenitrogenation, is a reaction between an organic nitrogen compound and hydrogen in which ammonia is released. This reaction usually takes place during the hydrodesulfurization of petroleum fractions and coal liquids. It is especially important for removing nitrogen compounds in the naphtha feed to a catalytic reformer since the catalyst is severely poisoned by all types of nitrogen compounds.

Catalyst poisoning is due to the neutralization of the acid sites of the alumina support. Acid sites — as will be explained later (Section 3.2.3 Cat-reforming) are responsible for carbonium ion formation which promotes isomerization and cyclization reactions.

Nitrogen compounds are mainly heterocyclic, having an aromatic character. The hydrogenolysis of these compounds appears to start by the saturation of the heterocyclic ring. The ring is then cleaved to yield primary and secondary amines. The amines further cleave to give ammonia and the corresponding hydrocarbon. The following are typical examples of the hydrogenolysis of some nitrogen compounds.

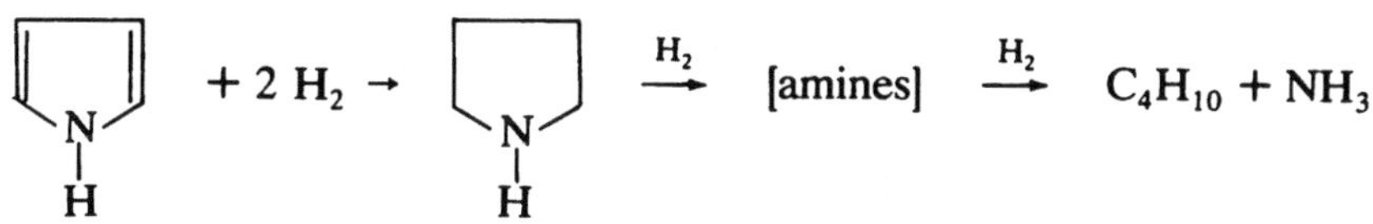

Pyrole

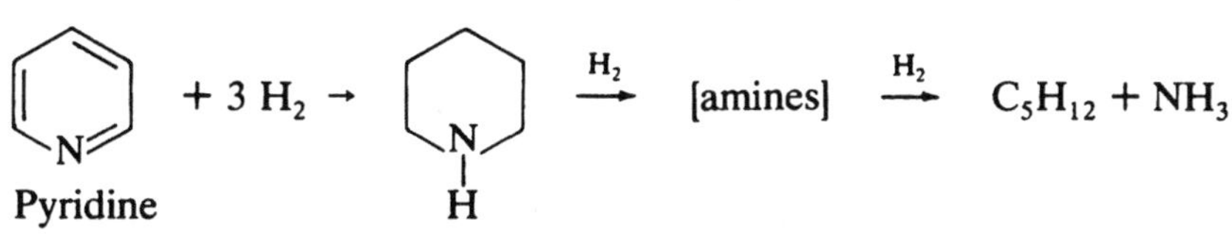

Pyridine

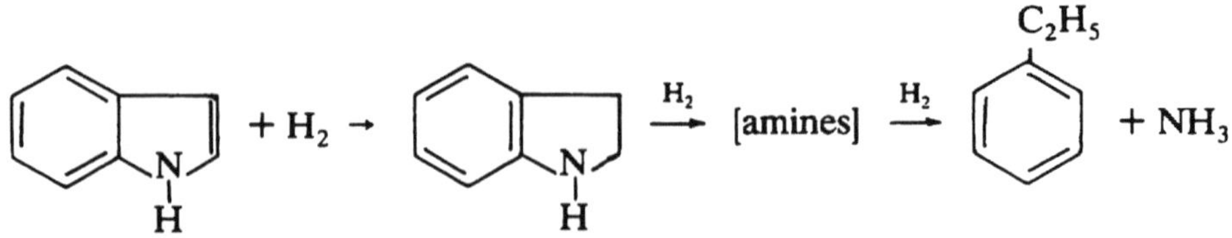

Indole

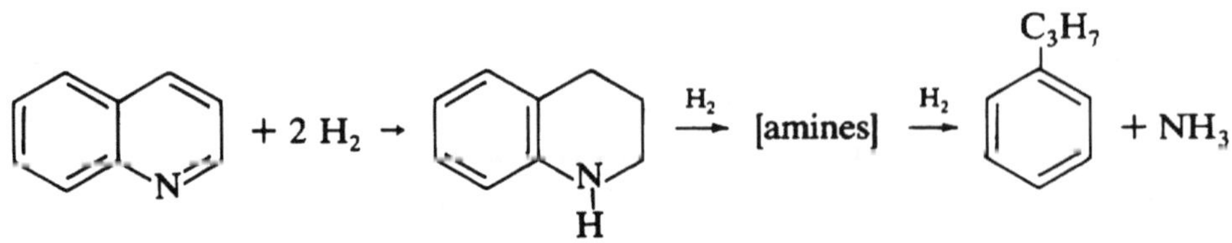

Quinoline

Catalysts for hydrodenitrogenation are similar to those used for hydrodesulfurization; the sulfide forms of the catalysts being more reactive than the oxide types.

3.1.2.5. *Hydrocracking*

Hydrocracking is a catalytic cracking process that takes place in the presence of hydrogen. It is aimed at converting high boiling fractions and residues to lower boiling components such as naphtha and jet fuels. Hydrocracking is particularly suitable for processing low value stocks which are not suitable for a catalytic cracking unit, because of the high metal, sulfur and/or nitrogen content. Cracking of long chain compounds and hydrogenation of the cracked products are the main reactions that take place. The hydrocracking reaction could be simply represented as:—

$$R{-}CH_2{-}CH_2{-}R' + H_2 \rightarrow R{-}CH_3 + R'{-}CH_3$$

In general, higher molecular-weight compounds crack more easily than lower molecular weight components. It is also known that complex aromatic compounds are the hardest of all hydrocarbon classes to crack. Other reactions, such as hydroisomerization, hydrodealkylation and hydrodesulfurization take place in a hydrocracking unit.

From low value feed stocks come a wide variety of products which include naphtha, jet fuel (kerosine) middle distilates, lubricating oils etc.

Dual function catalysts are used in hydrocracking processes. These catalysts are characterized by providing cracking sites and hydrogenation/dehydrogenation sites. The hydrogenation/dehydrogenation components are usually cobalt, nickel, chromium, tungsten, vanadium or their oxides. These catalysts are usually presulfided before operation and the combination of two or more metals is a normal practice.

Cracking sites, on the other hand, are provided by silica/alumina combinations or by zeolites. Zeolites not only possess higher activity than amorphous silica/alumina but they also have a better tolerance toward nitrogen compounds and thus a lower activation rate.

The choice of a catalyst depends on the required products and on feed specifications. Catalysts with strong acid sites promote cracking and isomerization reactions. A catalyst with a high cracking activity increases the gas yield. Isomerization is especially important when maximizing gasoline production, since iso-paraffins have higher octane ratings than *n*-paraffins. A high hydrogenation activity will reduce aromatics and olefins production. However, olefins are usually produced from a hydrocracking unit in only low yield.

Hydrocracking units could be adjusted to maximize certain products such as gasoline, jet or diesel fuels. A modification of the feed structure, the acidity of the catalyst, the conversion levels and the cracking temperature, could maximize one product over the other. For example, with a silica-rich catalyst, the amount of gasoline could be increased at the expense of diesel oil. Table 3-1 shows the

Table 3-1. Product analysis from a two stage hydrocracking unit. (Feed: light vacuum distilate 350–550).

Characteristics	First Stage	Maximum Jet Fuel	Maximum Diesel Oil
Yield, wt % feed			
$H_2S + NH_3$	2.20	2.20	2.20
$C_1 + C_2$	0.50	0.60	0.58
$C_3 + C_4$	1.40	8.77	3.40
Light gasoline	2.40	14.09	7.48
Heavy gasoline	3.20	16.92	13.50
Jet fuel	16.50	60.52	...
Diesel oil	25.60	...	75.36
Residue	50.00		
H_2 consumption, norm cu. m/ton	201	347	282
Products			
Jet fuel			
Specific gravity		0.795	
Freezing point, °C		−60	
Smoke point, mm		25	
Sulfur, ppm		5	
Diesel oil			
Specific gravity			0.820
Cetane number			58
Pour point, °C			−30
Sulfur, ppm			10
Residue			
Specific gravity	0.840		
Viscosity, cS at 210 °F	4.25		
Sulfur, wt %	0.001		
Nitrogen, ppm	1		

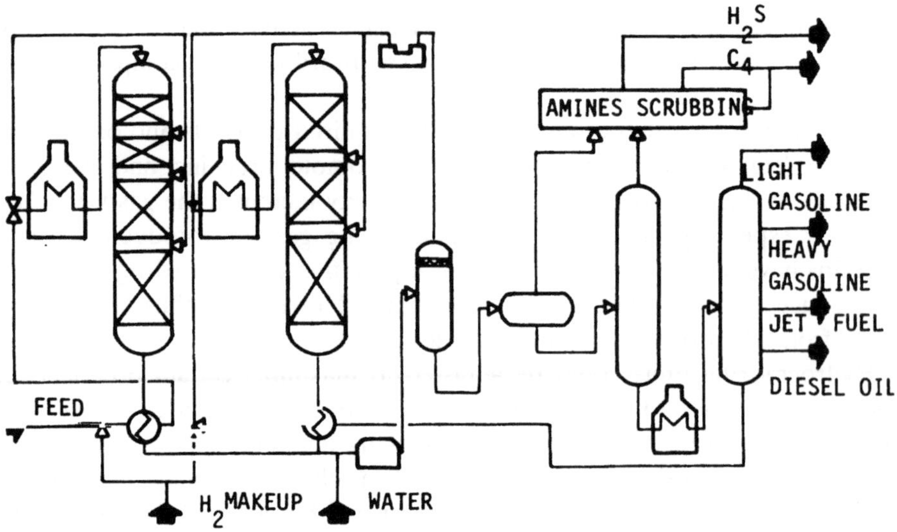

Fig. 3-1. A two stage hydrocracking unit.

analysis of feed and products from a hydrocracking unit, in one way to maximize gasoline and jet fuel, and in another to maximize diesel oil.

Hydrocracking could be a one or a two stage process. In the two stage hydrocracking the feed is hydrodesulfurized in the first stage, followed by distillation. The residue from distillation is then hydrocracked in a second stage (Fig. 3-1). Products from a one stage hydrocracking unit may require further treatment by hydrodesulfurization.

3.1.2.6. *Hydrodealkylation (of Toluene)*

Hydrodealkylation is a reaction between an alkyl benzene and hydrogen and can take place by a purely thermal process at 550—800°C. The carbon—carbon bond between the benzene ring and the alkyl group breaks and reacts with hydrogen giving a hydrocarbon and benzene. This reaction normally takes place but not to an appreciable extent during naphtha reforming. Since benzene is a more valuable petrochemical intermediate than toluene, a catalytic process — toluene hydrodealkylation — has been developed to increase benzene production.

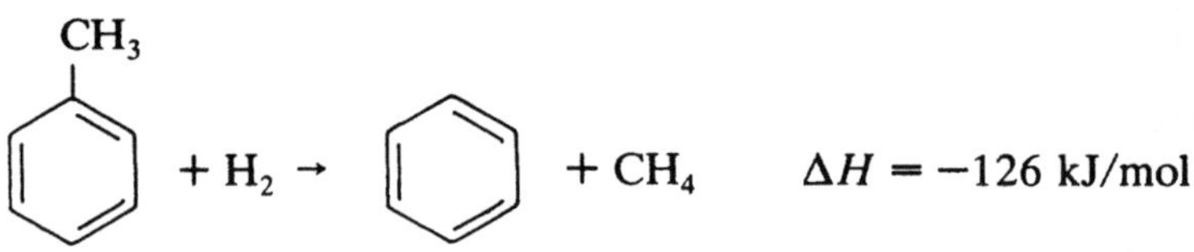

A reactor temperature ranging between 500—650°C and a pressure of about 50 bar are typical for a hydrodealkylation unit. Catalysts used are the oxides of chromium, molybdenum and cobalt on a support. When toluene is the only feed to the hydrodealkylation unit, one mole of hydrogen is consumed per mole of toluene. However, if the feed includes xylenes, the hydrogen consumption will depend on the mole ratio of the charge. Since methane is produced in this process, recycle is usually subjected to a low temperature process to separate methane from hydrogen. Product methane may be steam reformed to produce the required hydrogen for the hydrodealkylation unit.

$$CH_4 + H_2O \rightarrow CO + 3\,H_2$$

3.1.2.7. *Hydrogenation of Coal and Coal Liquids (Direct Coal Liquefaction)*

Coal is a combustible rock composed mainly of an organic material mixed with some inorganic compounds. Classifying coals into different ranks depends on the degree of coalification (time and conditions under which the original matter has been converted to coal). The highest ranked coal, anthrathite, has the lowest H/C ratio while lignite — the least coalified — has the highest H/C ratio of all coal types.

Low-ranked coals such as lignite and bituminous coals may be used as a source of a liquid fuel (synthetic crude) by raising their H/C ratio to a level comparable to that of some crude oils. This could be done by a noncatalytic or a catalytic hydrogenation process.

Catalysts used for the hydroliquefaction of coals are similar to those used for hydrodesulfurization of petroleum resids. Iron oxide mixed with alumina and molybdenum oxides are examples of these catalysts. Pressures in the range of 200—250 bar and temperatures in the range of 400—450°C are normally used to affect the liquefaction. Lower pressures and short contact times produce heavier liquids. Higher temperatures and long residence times reduce liquid yield and increase the amount of gases produced.

Another version of catalytic liquefaction of coal is the solvent extraction process. A hydrogen-rich solvent, which may be a coal-derived liquid, transfers hydrogen to the hot coal. A pressure of about 200 bar and a temperature of about 400°C are used. A catalyst may be used or, otherwise, the coal ash, which contains inorganic salts, can catalyze the liquefaction reactions.

In the 'Hydrogen-donor Process' a hydrogen carrier such as tetralin is used. It is capable of giving up its hydrogen to the decomposing coal molecules while being reduced to dihydronaphthalene.

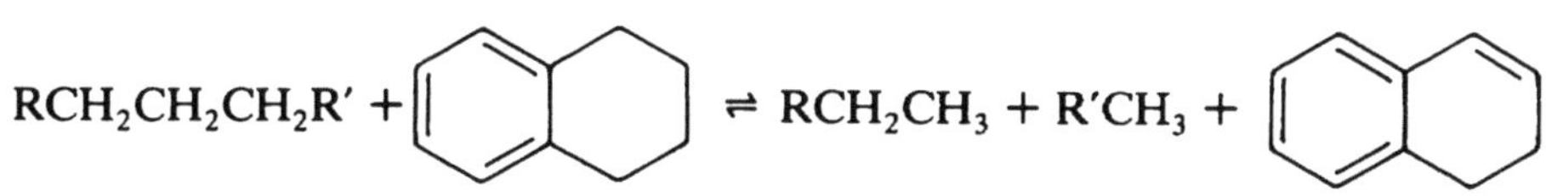

The solvent is recycled, when the dihydronaphthalene becomes rehydrogenated to tetralin.

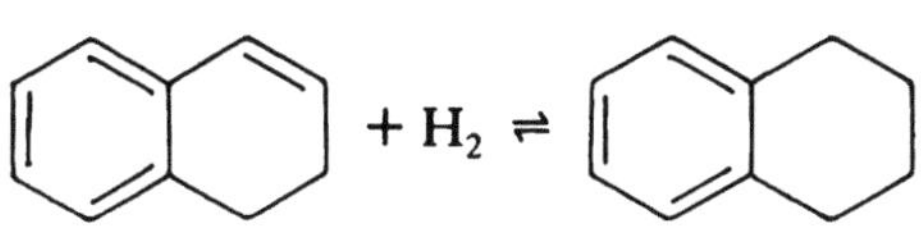

A catalyst system made of molybdenum and cobalt oxides supported on a highly porous alumina is typical. Other oxides of metals or their mixtures such as nickel, iron, and tungsten could also be used.

During the hydroliquefaction of coals, hydrodesulfurization and hydrodenitrogenation reactions take place. An advantage of the use of tetralin in the hydrogen-donor process is the highly effective hydrogen concentration that can be attained. One barrel of tetralin is equivalent to about 1850 scf of available hydrogen. Using hydrogen directly to reach a comparable concentration to that obtained with tetralin would require very high pressures. It is expected, then, that hydrogenation of sulfur compounds present in coals could be affected at much lower pressures. If high pressures are used in a donor-solvent liquefaction process, however, reactions are then expected to proceed at faster rates.

3.2. Dehydrogenation

In contrast to hydrogenation, dehydrogenation reactions are endothermic and are favored at higher temperatures. Since hydrogen is a product, dehydrogenation reactions are also favored at lower hydrogen partial pressures. In dehydrogenation, the number of product moles is always larger than the number of reactant moles and accordingly lower operating pressures shift the equilibrium towards the dehydrogenated products.

Rates of dehydrogenation reactions not only depend on the operating temperature, but also on the type of compound or compounds being dehydrogenated. In general C-6 cycloalkanes dehydrogenate much faster than the corresponding alkanes.

Major applications of catalytic dehydrogenation in the petroleum and petrochemical field are:—

- Dehydrogenation of C-4 and C-5 alkanes and alkenes for the production of the corresponding diolefins (butadiene and isoprene).
- Dehydrogenation at C-6—C-18 alkanes to monoalkenes.
- Dehydrogenation of naphtha (composed mainly of C-6$^+$ hydrocarbon mixture).
- Dehydrogenation of ethylbenzene.
- Dehydrogenation of alcohols.

3.2.1. DEHYDROGENATION OF C-4 AND C-5 ALKANES AND ALKENES

Production of Butadiene

1,3-Butadiene is an important monomer for the production of synthetic rubber. Its importance stems from the ease with which it can be polymerized and copolymerized. When it is polymerized, polybutadiene is produced. Copolymerization of 1,3-butadiene with styrene produces styrene-butadiene rubber (SBR) which is considered to be one of the most important synthetic rubbers for tire production.

Nitrile rubber (NBR) is produced by copolymerizing acrylonitrile with butadiene. A major use of nitrile rubber is for NBR latex. The copolymer of butadiene with styrene and acrylonitrile (ABS) is an important plastic which has special mechanical properties useful for engineering purposes, as compared with natural and other synthetic rubbers.

Although most of the butadiene needed comes as a co-product from ethane cracking, a substantial amount is synthesized by the catalytic dehydrogenation of butane, butenes, or their mixture.

$$2\ CH_3{-}CH_2{-}CH_2{-}CH_3 \rightarrow CH_3{-}CH{=}CH{-}CH_3 + CH_2{=}CH{-}CH_2{-}CH_3 + 2\ H_2 \qquad \Delta H = +126\ \text{kJ/mol}$$

$$\text{Butene mixture} \rightarrow CH_2{=}CH{-}CH{=}CH_2 + H_2 \qquad \Delta H = +109\ \text{kJ/mol}$$

The dehydrogenation reaction is favored by lowering the hydrocarbon partial pressure. This can be achieved by using steam, which has the additional advantage of decreasing coke formation on the catalyst surface. The reactor operates at approximately 650°C and 1 bar when butene is the feed. Conversion usually does not exceed 50% with about 90% selectivity for butadiene. The catalyst used is calcium-nickel phosphate impregnated with chromium oxide. Other processes do not use steam but are operated at lower pressures. The catalysts generally used are based on alumina impregnated with chromic oxide, iron oxide and metal halides.

These catalysts are sensitive and deactivate fast in presence of steam.

In the butane operation (more endothermic) a higher temperature (about 900°C) than in butene dehydrogenation is required. This promotes other reactions which are activated at higher temperatures such as cracking and polymerization. Short residence times increase the selectivity toward butadiene.

Oxidative dehydrogenation of *n*-butenes at a relatively lower temperature range (480—600°C) is highly selective for butadiene at high conversion (about 90% selectivity at 80% conversion).

$$CH_3{-}CH_2{-}CH{=}CH_2 + CH_3{-}CH{=}CH{-}CH_3 + O_2 \rightarrow 2\ CH_2{=}CH{-}CH{=}CH_2 + 2\ H_2O$$

The effect of oxygen/*n*-butene ratio on conversion and yield is shown in Fig. 3-2. A typical analysis of feed, reactor effluent and crude butadiene from pilot unit operation is shown in Table 3-2.

Production of Isoprene (*2-methyl-1,3-butadiene*)

Isoprene is also an important monomer for synthetic rubber production. Stereoselective polymerization of isoprene produces *cis*-polyisoprene which has a similar structure and similar properties to natural rubber. Copolymerization of isoprene with isobutene produces butyl rubber (BR). Isoprene is a co-product with ethylene

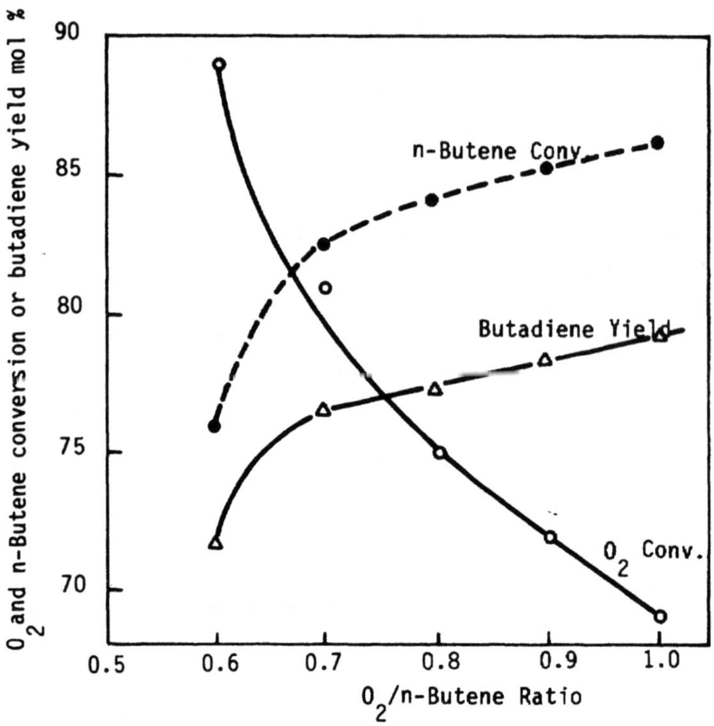

Fig. 3-2. Effect of O_2/*n*-butene ratio on conversion and yield (contact time: 1.8 s, *n*-butene conc.: 10 mol%).

from naphtha cracking. It is isolated from the C-5 fraction either by extraction or by distillation. It is also produced via different synthetic routes. An alternative method developed by Shell is dehydrogenating a C-5 olefin mixture. The mixture, 2-methyl-1-butene and 2-methyl-2-butene, is extracted with sulfuric acid from a C-5 cat-cracking fraction. The acid isomerizes 2-methyl-1-butene to 2-methyl-2-butene during the extraction step. It is then dehydrogenated by the use of a catalyst in a similar way to that used for butadiene production.

$$CH_3{-}\underset{}{\overset{CH_3}{\overset{|}{C}}}{=}CH{-}CH_3 \rightarrow CH_2{=}\overset{CH_3}{\overset{|}{C}}{-}CH{=}CH_2 + H_2$$

3.2.2. DEHYDROGENATION OF HIGHER ALKANES TO MONOALKENES

Higher monoalkenes (monoolefins) are produced by four different methods.

- Dehydrogenation of linear alkanes (*n*-paraffins).
- Oligomerization of ethylene.
- Dehydrochlorination of monochloroalkanes.
- Wax cracking.

The dehydrogenation route could be affected by reacting *n*-alkanes in the range of

Table 3-2. Typical analysis of feed, reactor effluent and crude butadiene from a pilot plant operation.

Component mol %	Reactor feed	Reactor effluent	Crude butadiene
Nitrogen	72.84	68.99	0.05
Oxygen	8.97	2.28	0.01
Argon	0.87	0.82	0
Carbon dioxide	1.50	2.64	0.53
Carbon monoxide	0.47	0.75	0
Water	—	10.32	0.11
Isobutane	1.29	1.20	8.59
n-Butane	3.88	3.61	25.99
Butene-1	4.47	0.17	1.23
trans-Butene-2	3.57	0.73	5.23
cis-Butene-2	2.13	0.45	3.20
Butadiene	0	7.52	53.89
Formaldehyde	—	0.10	0
Acetaldehyde	—	0.08	0
Furan	—	0.12	0.77
Acetone	—	(73 ppm)	0
Acrolein	—	0.09	0.02
Methacrolein	—	0.01	0.01
Benzene	—	0.03	0.17
Methylvinylketone	—	0.03	0.01
C_4 extractive solvent	—	—	0.19
Acetic acid	—	0.03	0
Isobutyric acid	—	t	0
Acrylic acid	—	0.02	0
Methacrylic acid	—	t	0
Methylacetylene	—	(2 ppm)	(16 ppm)
Ethylacetylene	—	—	0
Vinylacetylene	—	(12 ppm)	(85 ppm)
Butadiene dimer	—	—	(20 ppm)
Peroxides	—	—	(1.4 ppm)

C-6—C-18 over a dehydrogenation catalyst. A fixed-bed system is used at a temperature range of 400—600°C and high purity hydrogen is separated while a portion is recycled. Operating at low space velocity decreases the selectivity to monoalkenes since a high residence time allows competing reactions to take place. Conversion is low at very high space velocities (short residence times). To obtain a reasonable conversion at high space velocities, higher temperatures are used. This will also reduce the selectivity to monoalkenes due to increased rates of side reactions. Fig. 3-3 shows the effect of changing the space velocity on the selectivity to linear monoalkenes. Linear monoalkenes in the range of C-12—C-14 are used to alkylate benzene for the production of linear alkylbenzene.

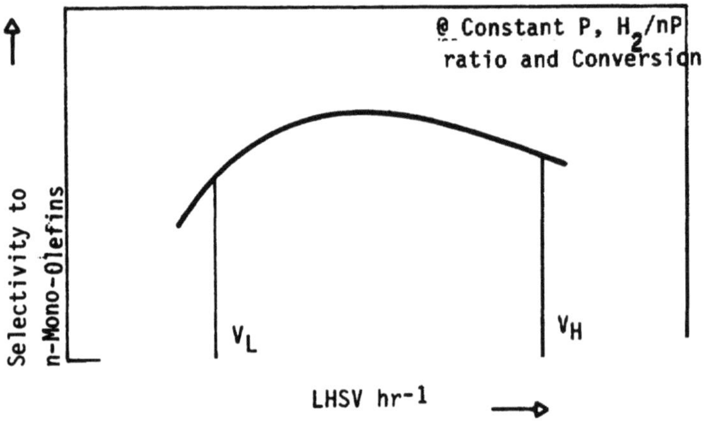

Fig. 3-3. Effect of changing space velocity on selectivity to monoalkenes.

3.2.3. CATALYTIC REFORMING OF NAPHTHA (PRODUCTION OF BENZENE, TOLUENE AND XYLENES)

Catalytic naphtha reforming is essentially a dehydrogenation process. However, other important reactions such as isomerization, hydrocracking, and hydrodealkylation take place during the reforming process. These reactions will be discussed briefly since they greatly influence the reforming process variables and the product composition.

Historically, naphtha reforming was developed to obtain a high octane gasoline by thermally changing the molecular structure of the hydrocarbon mixture. Catalytic naphtha reforming soon replaced the thermal process for the same purpose. Catalysts then in use were metal oxides, which were unstable. Platforming was first introduced by UOP, when it was discovered that platinum is more stable and active than catalysts previously used. Bi-metallic and multi-metallic catalysts developed later provided even more stability and activity than monometallic platinum catalysts. However, all these catalysts contain platinum with one or more metals, such as rehnium. The commercial process using Pt/Re is called Rheniforming.

Depending on the feed and reactor conditions, catalytic reformers are currently designed to produce reformates of 100 or more Research Octane Number with a composition of up to 70 volume percent aromatics from a naphtha feed having about 10 volume percent aromatics.

Originally, all aromatics for chemical use were made from coal. Currently about 5% of the total demand for aromatics derives from coal. The rest comes from catalytic reforming, catalytic cracking, and from disproportionation and hydrodealkylation of toluene. Benzene is also a by-product of ethylene cracking. Catalytic crackers produce cracked gasoline which contains only about 22 volume percent aromatics. The composition of cracked gasoline is however relatively hard to control. Sources for benzene, toluene, and xylenes (BTX) may also change with the development of coal liquefaction. Coal liquids are highly aromatic in nature and may constitute a substantial source for BTX by the approach of the twenty-first century.

The importance of BTX for the petrochemical industries is related to their activities as chemical compounds that can be alkylated, oxidized, or hydrogenated to give a great number of useful chemicals and intermediates.

Reforming serves the dual purpose of producing both high octane gasoline and aromatics for chemical use. The octane rating of reformed naphtha is a function of the proportion of different hydrocarbons in the reformate and their respective octane numbers. It is known that aromatic compounds and branched alkanes are responsible for the high octane rating of reformed gasoline. The tendency is toward maximizing aromatics for their usefulness as petrochemical feedstocks. This is limited, however, by the catalyst type, the feed and operation conditions. The effect of these variables on BTX yield is discussed.

Catalysts

The first catalysts used in catalytic reforming were chromium oxide or molybdenum oxide supported on alumina.

These catalysts were unstable and produced large amounts of coke which deactivated the catalysts after short periods. UOP was the first to introduce a platinum-alumina catalyst in 1949. This catalyst system proved more active and selective than the oxides previously used. Platforming is a naphtha reforming process using 0.3—0.7 weight percent platinum on an alumina carrier. In catalytic reforming two catalytic sites are needed: hydrogenation—dehydrogenation sites, and acid sites which promote reactions by a carbonium ion intermediate. Platinum is the best known hydrogenation—dehydrogenation catalyst. Acid sites are provided by the alumina carrier. Both these sites are necessary for aromatization and isomerization reactions.

To improve the catalytic activity and selectivity for aromatics, bimetallic and multimetallic catalysts have been extensively investigated. A bimetallic catalyst of Pt/Re or a tri-metallic catalyst of a noble metal alloy provides the hydrogenation—dehydrogenation sites. The increased stability of these bi- and multi-metallic catalysts allowed reforming of naphtha under lower pressures, thus increasing BTX production. Operating at lower pressures has the additional advantage of reducing compression costs. Reforming at lower pressures, however, produces coke which deactivates the catalyst rapidly, and shortens the regeneration cycles.

Aromatization Reactions

Aromatization reactions are those leading to the formation of an aromatic compound from a nonaromatic hydrocarbon.

Important aromatization reactions are the dehydrogenation of naphthenes (saturated cyclic hydrocarbons) and the dehydrocyclization reaction of alkanes. The first is fast and reaches equilibrium in a short time. The reaction is a dehydrogenation of cyclohexane or a substituted cyclohexane on the dehydrogenation sites of a metal such as platinum. Substituted cyclopentanes require an additional step, isomerization to cyclohexane followed by dehydrogenation. Equation (1) illustrates the dehydrogenation of cyclohexane to benzene.

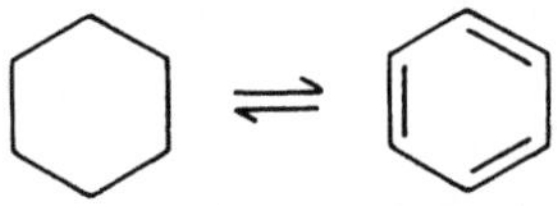

$$+ 3\,H_2, \qquad \Delta H = 221 \text{ kJ/mol} \tag{1}$$

The reaction is reversible and is highly endothermic. From the balanced equation, the equilibrium constant is 6×10^5 at 500°C (pressure in atmospheres).

$$K_p = \frac{P_{Ar} \cdot P^3_{H_2}}{P_{naph.}} = 6 \times 10^5 \text{ @ } 500°\text{C}$$

It is clear that the yield of benzene is favored at higher temperatures and lower pressures. The effect of pressure varies dramatically with the change in H_2 partial pressure. Reforming units are normally run within the temperature and pressure ranges 500—530°C and 14—25 kg/cm^2 respectively.

The second important reaction is dehydrocyclizing alkanes. Dehydrocyclization of alkanes involves consecutive reactions. The first step is dehydrogenation to an alkene. Other steps follow which may involve the formation of a carbonium ion. Using *n*-hexane as an example, the reaction could be represented by Equation (2)

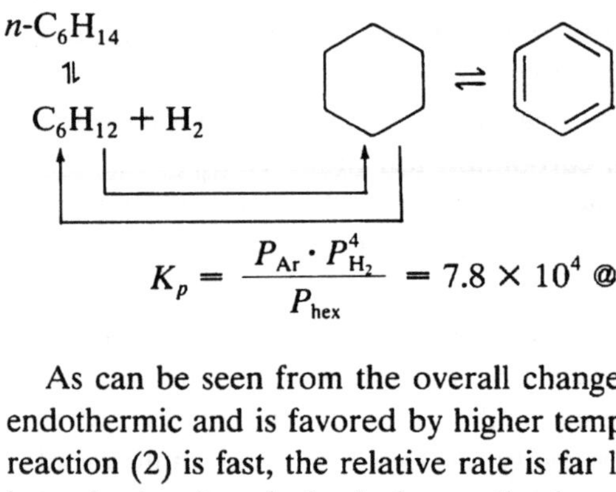

$$+ 3\,H_2 \qquad \Delta H = +266 \text{ kJ/mol} \tag{2}$$

$$K_p = \frac{P_{Ar} \cdot P^4_{H_2}}{P_{hex}} = 7.8 \times 10^4 \text{ @ } 500°\text{C}$$

As can be seen from the overall change in enthalpy, this reaction is also highly endothermic and is favored by higher temperatures and lower pressures. Although reaction (2) is fast, the relative rate is far less than that of reaction (1). This could be understood on the basis that cyclization of hexane to cyclohexane is slow.

Table 3-3 shows the result of reforming hexane over a platinum catalyst at

Table 3-3. Results of reforming *n*-hexane over platinum catalyst at LHSV 2 h^{-1}.

Temp., °F	% Conversion	Selectivity to benzene	Selectivity to isohexane
885	80.2	16.6	58
932	86.8	24.1	36.9
977	90.4	27.4	23.4

different temperatures. Cyclization can take place by carbonium ion attack on a carbon—carbon double bond in the same molecule. For example, toluene can be produced by the following reaction sequence.

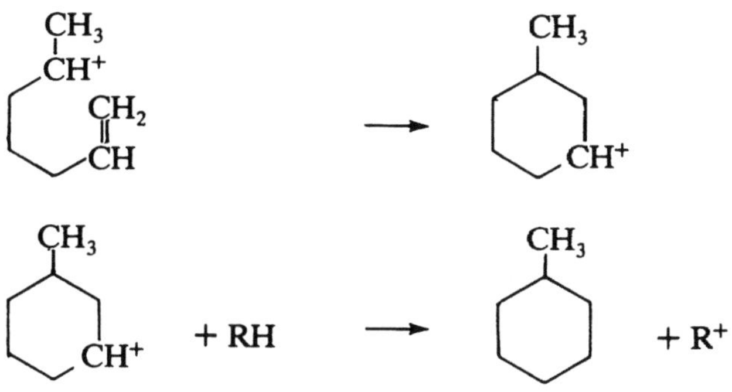

Methylcyclohexane can dehydrogenate on the platinum surface by adsorption on a hydrogenation–dehydrogenation site. Another alternative for the aromatization step is via a hydrogen transfer reaction of the cyclic carbonium ion intermediate.

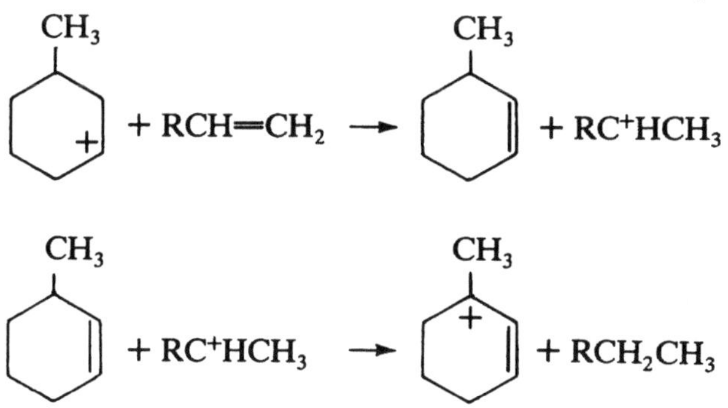

A continuation of this sequence converts the cyclic carbonium ion to toluene.

An important reaction that is indirectly related to aromatization is the isomerization of alkanes and cycloalkanes. Isomerization may take place on the surface of a non-acidic catalyst such as platinum or non-acidic alumina or by the aid of an acid catalyst. In the former case the reaction is slow. It has been found that when 2-methyl-pentane was reformed using Pt on non-acidic alumina, benzene appeared as a secondary product. The dehydrocyclization of hexane may take place either directly on the catalyst surface or through the formation of a five-membered ring intermediate which isomerizes to cyclohexane. It seems that 2- and 3-methylpentanes may have to isomerize to *n*-hexane or alternatively may form a five-membered ring intermediate before dehydrogenation to benzene. This sequence of reactions is shown in Equation (3).

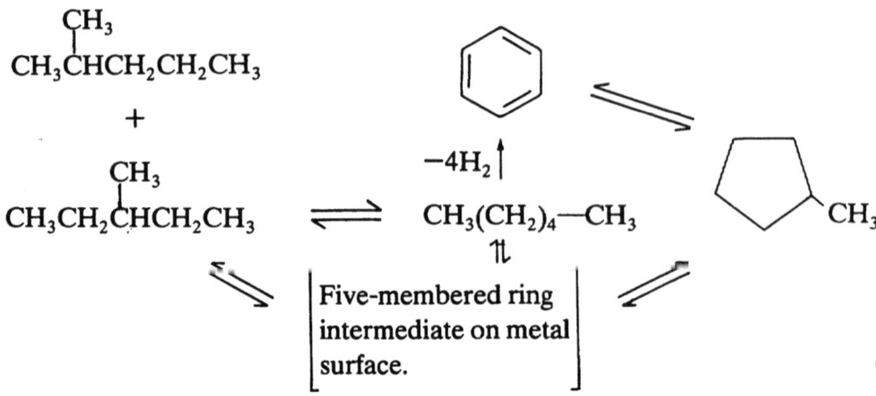

(3)

Acidic catalysts, on the other hand, catalyze the isomerization reaction much faster than non-acidic catalysts, through the intermediate formation of a carbonium ion. The isomerization step may be more significant for hydrocarbons having fewer than six carbons in the main chain such as methylpentanes and dimethylbutane. These compounds can isomerize to *n*-hexane which is able to dehydrocyclize to benzene. When acid catalysts are used it seems that the dehydrogenation of *n*-hexane to an *n*-hexene would be the first step. The alkene is then protonated to a carbonium ion intermediate followed by cyclization. Whether cyclization takes place through a methylcyclopentane intermediate followed by isomerization to cyclohexane, or by direct cyclization, is not clear.

The SRI evaluation of reforming for BTX product composition was not successful until it was realized that the concept that ring closure has to be to cyclopentane rather than to cyclohexane had to be adopted.

In this discussion, *n*-hexane and cyclohexane have been used as examples to represent the main aromatization reactions. However, naphtha feed to a reformer is a mixture of hydrocarbons with different chain length and with many isomers. The rates of the reactions, as well as the equilibrium, may be greatly affected by the different hydrocarbon concentrations in this reactant mixture. Relative rates of reforming reactions for different hydrocarbons are shown in Table 3-4.

Table 3-4. Relative rates of major reforming reactions.

Reaction	Relative rate
Paraffins → Naphthenes	1
Paraffins → Cracked products	1
Naphthenes → Aromatics	32.5
Naphthenes → Cracked products	1
Naphthenes → Paraffins	6.5
Aromatics → Naphthenes	1

Acid sites of the catalyst carrier are thus important for isomerizing substituted cyclopentanes to cyclohexanes before dehydrogenation to aromatics. These sites also catalyze the isomerization of branched C-4 and C-5 alkanes to hexanes before being able to cyclize to naphthenes. The acid carrier is also effective for the isomerization of substituted cyclohexanes and aromatics. As a result, C-8 aromatics are always produced very close to their thermodynamic equilibrium values.

Both isomerization and aromatization reactions are important for raising the octane rating of product gasoline. For petrochemical use, however, reforming conditions are set to favor aromatization for maximizing BTX. Figure 3-4 illustrates selectivities of different hydrocarbons in a naphtha feed for aromatization in a catalytic reformer.

Other Important Reforming Reactions

Dealkylation. Dealkylation of the aromatic side chain takes place under certain favorable conditions. This reaction is particularly important for increasing benzene yield when toluene and ethylbenzene are dealkylated.

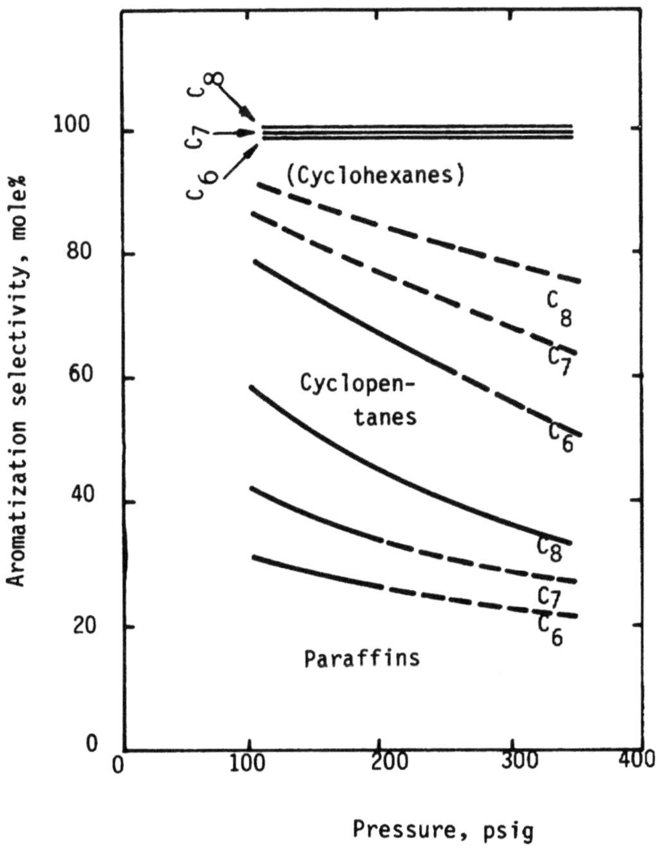

Fig. 3-4. Aromatization selectivities in catalytic reforming.

Using toluene as an example, the reaction is slightly exothermic and is favored at a relatively lower temperature.

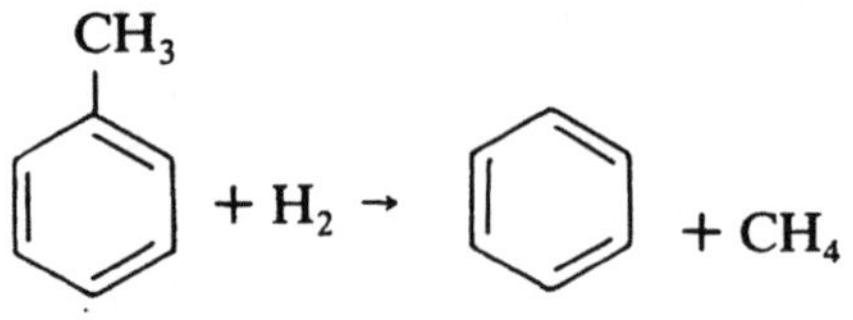

A higher benzene yield is expected from this reaction at higher hydrogen partial pressure. Dealkylation is not limited to toluene and ethylbenzene, but also polyalkylated benzenes such as xylenes and trimethylbenzenes are susceptible to this reaction, downshifting to the BTX range. For example, 1,2,4-trimethylbenzene may be dealkylated to xylenes, which can further be dealkylated to toluene.

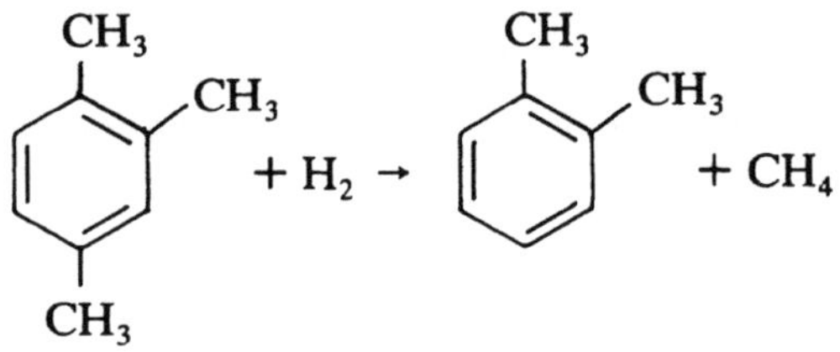

Disproportionation. Disproportionation of toluene to benzene and xylenes takes place during reforming. This is an equilibrium reaction which depends on the ratio of the three components, benzene, toluene, and xylenes in the reaction mixture. The reverse reaction is the transalkylation of benzene and xylenes to give toluene.

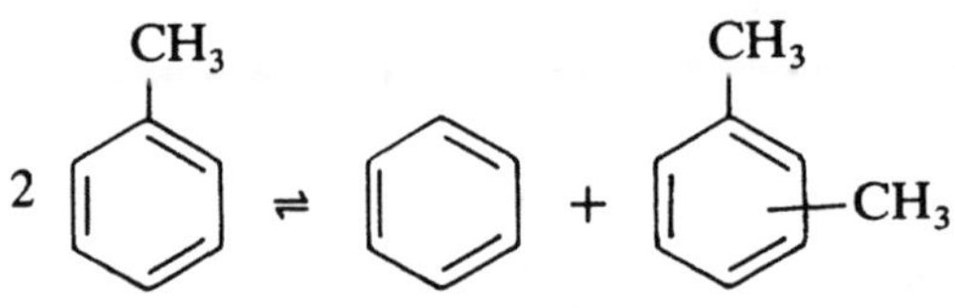

Due to the need for benzene for downstream petrochemicals production, catalytic hydrodealkylation of toluene is used for this purpose. It is also disproportionated to benzene and xylene with a theoretical conversion of 58%. Typical conversions in the order of 40% are achieved without many side reactions.

Hydrocracking. Hydrocracking is a cracking reaction in the presence of hydrogen. It is responsible for a high gas yield in naphtha reforming. The reaction is favored at high temperatures and under high hydrogen partial pressure. It can be represented as:

$$R{-}CH_2{-}CH_2{-}CH_2{-}R' + H_2 \rightarrow RCH_3 + R'{-}CH_2{-}CH_3$$

Bond breaking may occur at any place along the hydrocarbon chain producing two hydrocarbon molecules. It is generally accepted that longer chains crack more readily than shorter chain hydrocarbons. This may be important when reforming higher molecular weight naphthas that will crack to C-6, C-7, and C-8 hydrocarbons which are suitable for dehydrocyclization to the corresponding aromatics.

Naphthenes crack faster than paraffinic hydrocarbons. However, their amounts in the reaction mixture might not be sufficient to contribute much to this reaction. This is because the rate of the reaction of naphthenes to give aromatics is much faster than that to give cracked products. For a paraffinic naphtha feed, the end point may be increased to favor hydrocracking and dehydrocyclization reactions for more BTX production. However, increased hydrocracking results in liquid product losses due to the excessive coproduction of hydrocarbon gases.

Feed Stocks. Naphtha feeds to a reformer differ in their composition depending on their origins. Feed naphthas rich in naphthenes will allow reformers to be run at lower severity than paraffinic naphthas. Reformates from highly naphthenic feeds will be rich in aromatics than from paraffinic feeds.

Another approach for increasing the yield of BTX aromatics is to reform naphthas with high end-points. This will promote the formation of higher alkyl benzenes, since more possibilities exist for closing the ring. The alkylbenzenes are further hydrodealkylated to BTX. However, it has been argued that this practice results in higher coke formation. Table 3-5 shows the analysis of an Arabian naphtha feed and the products from a reforming unit.

Reforming Process. Catalytic reformers (Fig. 3-5) are designed to have a series of catalyst beds. Since the rates of the different hydrocarbon reactions vary, the first bed contains a catalyst with lower platinum concentration. As shown in Table 3-4,

Table 3-5. Analysis of products from a catalytic reforming unit using an Arabian naphtha feed.

FEEDSTOCK	HYDROFINED ARABIAN NAPHTHA	
Boiling range, °F	189—310	
Composition, LV%		
Paraffins	68.6	
Naphthenes	23.4	
Aromatics	8.0	
Sulfur, ppm	0.2	
Nitrogen, ppm	0.5	
Reactor outlet pressure, psig.	100—200	
Product Yields		
Hydrogen, scf/bbl feed	1,530	1,260
C_1—C_3, scf/bbl feed	195	340
C_5^+ Reformate, LV%	79.1	74.1
Reformate Properties		
C_5^+ Research octane, clear	98	99
Composition, LV%		
Paraffins	32.2	31.7
Naphthenes	1.7	0.5
Aromatics	66.1	67.8

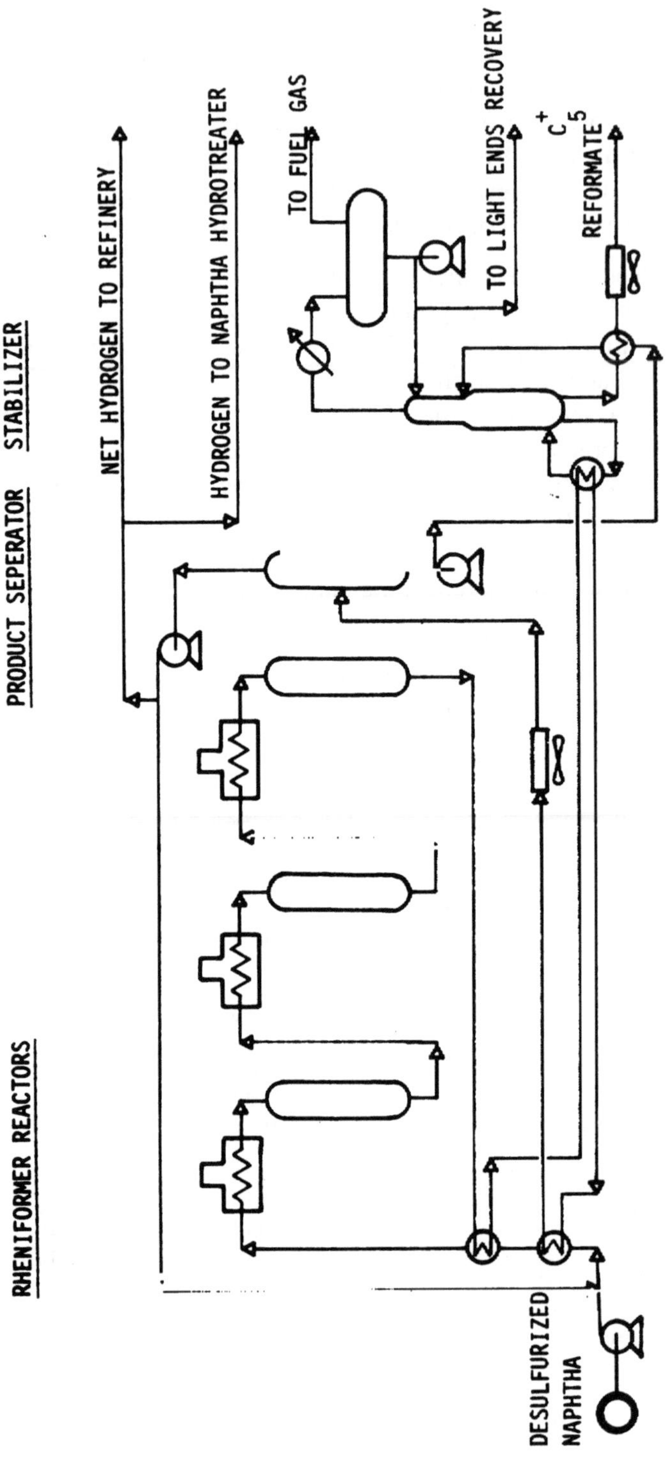

Fig. 3-5. A three-bed rheniforming process.

the relative rates of dehydrogenation of naphthenes, are very fast and may reach equilibrium in the first bed. Since the dehydrogenation reaction is highly endothermic, the reactor outlet temperature (about 430°C) is lower than the inlet temperature (about 520°C). The effluent from the first bed is reheated back to about 520°C before flowing to the next reactor. In the second and third beds, slower dehydrocyclization of paraffins, isomerization, disproportionation, and hydrocracking reactions take place. The first reaction is endothermic while the others are mildly exothermic. However, the overall heat of reaction is endothermic. Heat losses in the second and third reactors are less than in the first.

Liquid hourly space velocities in the range of 1—2 h^{-1} are typical. Higher space velocities are possible in the first reactor since dehydrogenation of naphthenes is fast. Lower space velocities favor hydrocracking activities.

The hydrogen partial pressure has a great effect on aromatization. Lower hydrogen partial pressures favor aromatics production. However, operating at lower hydrogen partial pressure promotes coke formation.

Catalyst regeneration depends on the reforming process. Reformers are built with cyclic semiregenerative or continuous regeneration systems. In the cyclic process, a swing reactor is used to replace any one of the reactors while it is regenerated. In the continuous system, the catalyst moves through a series of beds which passes to a regeneration system and then back to the reactor.

Both schemes allow for continuous operation with no loss of reforming capacity.

3.2.4. DEHYDROGENATION OF ETHYLBENZENE

Production of Styrene

Styrene is an important monomer used for the production of styrene polymers and copolymers. Polystyrene is the third largest thermoplastic after polyethylene and polyvinyl chloride. The copolymer with acrylonitrile (SAN) has a higher tensile strength than polystyrene alone. Copolymers of styrene and butadiene in a ratio of approximately 1 to 25 are the most important synthetic rubber (SBR). The main route to styrene is the catalytic vapor phase dehydrogenation of ethylbenzene. The reaction is highly endothermic and a temperature of approximately 600—700°C is required for the dehydrogenation. The heat needed for the reaction may be supplied by burning the product hydrogen.

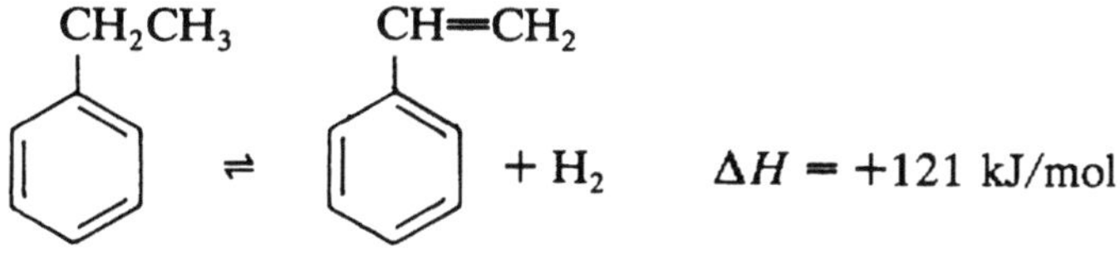

Many catalysts are used for the dehydrogenation of ethylbenzene. Oxides of iron, zinc, cobalt, silicon, or their mixtures are typical catalysts. Iron oxide impregnated with chromium oxide and potassium compounds produces styrene selectively (about 90%) with a higher conversion of ethylbenzene (approx. 70%) than older catalysts. Styrene is then separated from unreacted ethylbenzene and some by-products such as toluene and heavy tars. Raw styrene is then distilled under vacuum to minimize polymerization.

3.2.5. DEHYDROGENATION OF ALCOHOLS

Production of Aldehydes and Ketones

Dehydrogenation of alcohols, or in the chemical sense their oxidation to the corresponding aldehydes and ketones, is a well known reaction. Dehydrogenating a primary alcohol produces an aldehyde while dehydrogenating a secondary alcohol gives a ketone. Secondary alcohols are generally easier to dehydrogenate than primary alcohols. Catalytic dehydrogenation of primary alcohols is preferred, rather than oxidizing them by the use of oxidizing agents, since aldehydes can be further oxidized to the corresponding acids.

Catalysts used for dehydrogenating alcohols are silver, copper, zinc oxide and brass.

Methanol Dehydrogenation (Production of Formaldehyde)

Formaldehyde is the simplest and most reactive aldehyde. Condensation polymerization of formaldehyde with phenol, urea or melamine produces, respectively, phenol-formaldehyde, urea-formaldehyde or melamine-formaldehyde resins. These resins are of the thermosetting type. Formaldehyde condenses with acetaldehyde producing pentaerythritol, an important polyhydric alcohol. Its reaction with ammonia produces hexamethylenetetramine which is a cross-linking agent for phenolic resins.

The major route for the production of formaldehyde is by the dehydrogenation or oxidative dehydrogenation of methanol. Either silver alone in the form of crystals or nets is used, or else it is supported on carborundum. The first reaction releases hydrogen which could be utilized for providing the heat for the endothermic reaction.

$$CH_3OH \rightleftharpoons HCHO + H_2 \qquad \Delta H = +84 \text{ kJ/mol}$$

In the oxidative dehydrogenation, air as a source of oxygen is used to adiabatically supply the required heat, when it reacts with hydrogen giving water. Steam is usually added to lengthen the catalyst life.

$$H_2 + \tfrac{1}{2} O_2 \rightleftharpoons H_2O \qquad \Delta H = -243 \text{ kJ/mol}$$

Ethanol Dehydrogenation (Production of Acetaldehyde)

Acetaldehyde is an important precursor for acetic acid and acetic anhydride.

Currently, a major portion of acetic acid production is by the liquid phase catalytic oxidation of ethylene (Wacker catalysts). It may also be produced by the direct oxidation of ethane. Oxidative dehydrogenation of ethanol is also an important route to acetaldehyde. The exothermic oxidation provides the heat required for the endothermic dehydrogenation.

$$CH_3CH_2OH + \tfrac{1}{2}O_2 \rightleftharpoons CH_3CHO + H_2O \qquad \Delta H = -180 \text{ kJ/mol}$$

$$CH_3CH_2OH \rightleftharpoons CH_3CHO + H_2 \qquad \Delta H = 84 \text{ kJ/mol}$$

Isopropyl Alcohol Dehydrogenation (Production of Acetone)

Acetone (2-propanone) is the simplest and most important aliphatic ketone. As a solvent, it is mainly used in the paint and lacquer industry. In synthetic work acetone is used for the production of important chemicals such as methyl methacrylate, for acrylic fibers and bisphenol A for epoxy resins, polyurethane foams and polycarbonate plastics. Aldol condensation of acetone leads to the production of diacetone alcohol which the precursor of such important chemicals as mesityl oxide, methyl isobutylketone, and methyl isobutyl carbinol, which are industrial solvents for paints. Acetone is a co-product with phenol from cumene peroxidation, which is currently the major route for its production. Dehydrogenation of isopropanol is still an important process for producing acetone.

$$CH_3CHOHCH_3 \rightleftharpoons CH_3\overset{\overset{\displaystyle O}{\|}}{C}CH_3 + H_2 \qquad \Delta H = +67 \text{ kJ/mol}$$

The dehydrogenation reaction can take place in a gas phase process at a temperature range of 350–550°C using zinc oxide or a copper–zinc catalyst. A high conversion of isopropanol (approx. 98%) could be attained with a high selectivity to acetone. Dehydrogenation in the liquid phase using Raney nickel could be conducted at about 150°C.

References

Matar, S.: *Synfuels — Hydrocarbons of the Future*. PennWell Publishing Co., Tulsa, Oklahoma, 1982.

Hatch, L. and Matar, S.: *From Hydrocarbons to Petrochemicals*. Gulf Publishing Co., Houston, Texas, 1981.

Hagopian, C., Lewis, P., and McDonald, J.: 'Evaluate Styrene Production', *Hydrocarbon Processing* **62**(2) 1983, p. 45.

Wade, D., Ansell, L., and Epperly, W.: 'Coal Liquefaction', *CHEMTECH*, April 1982, p. 242.

Gates, B.: 'Liquefied Coal, by Hydrogenation', *CHEMTECH*, Feb. 1979, p. 99.

Hemler, C. and Wilcox, J.: 'Catalytic Cracking for the Production of Transportation Fuels from Syncrudes', American Petroleum Institute 46 mid-year Refining Meeting Part 1, Chicago, Illinois, 1981.

Billon, A., Frank, J., Peries, J., Fehr, E., Gallei, E., and Lorenz, E.: 'More Ways to Use Hydrocracking', *Hydrocarbon Processing* **57**(5) 1978, p. 122.

Yoshioka, A., Hokari, H., Sato, R., Harohisa, Y., and KInichi, O.: 'Make Butadiene from Butene/Butane Feed', *Hydrocarbon Processing* **63**(11) 1984, p. 97.

Matar, S.: 'Aromatics: Production and Chemicals', *The Arabian Journal for Science and Engineering* **11**(1), 1986.

'Rheniforming', *Hydrocarbon Processing* **63**(9) 1984, p. 119.

CHAPTER 4

Alkylation Processes

4. Introduction

The term alkylation has generally been used to describe the catalyzed reaction between isobutane and various light olefins. The product is a highly branched paraffinic hydrocarbon used as a blend to improve the octane number of gasoline. A more specific explanation of the term alkylation is the attachment of one or more alkyl groups to a hydrocarbon compound or to a hydrocarbon derivative. This may take place by an electrophilic attack of a carbonium ion species (whether formal or ion-pair) or by a nucleophlic attack of a carbanion or carbanion-like species. When a carbonium ion is the intermediate, alkylation can occur either by addition, as in the case of the reaction between a carbonium ion and an alkene, e.g.

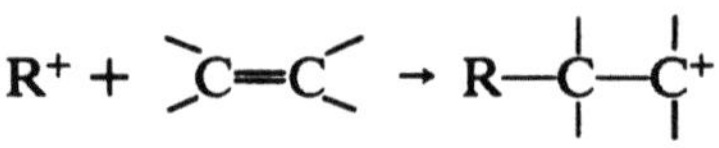

or by substitution as in the case of Friedel-Crafts type reactions e.g.

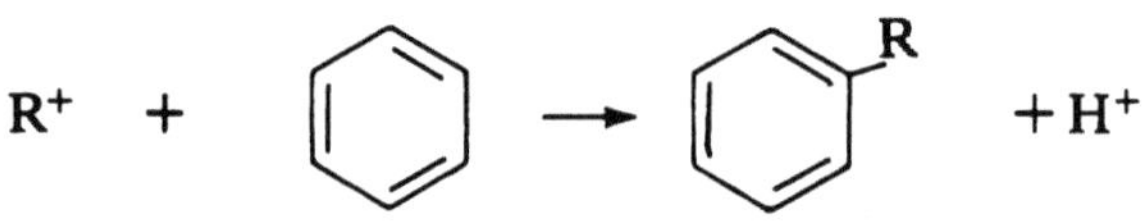

Examples of alkylations involving a carbanion or a carbanion-like species are Grignard reactions with aldehydes and ketones and malonic ester synthesis. In the former case the carbanion-like species adds to the carbonyl carbon and in the latter the carbanion acts as a nucleophile in a substitution reaction.

Alkyl groups can either be attached to a carbon atom, and the alkylation is a carbon alkylation, or to a heteroatom such as oxygen and, in this case, it is termed oxygen alkylation. Examples shown above are for carbon alkylations. Malonic ester synthesis can give either carbon or oxygen alkylation.

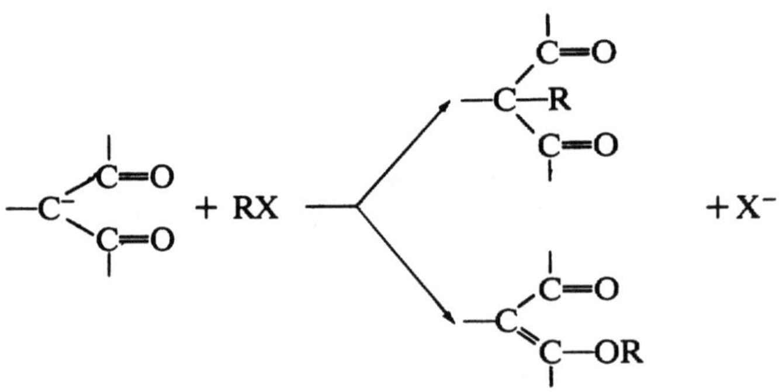

Alkylation reactions are useful in organic synthetic research as well as in industrial applications for the production of important chemicals. Examples of alkylates of commercial importance are ethylbenzene, cumene, linear alkylbenzene, alkylamines and 2,2,4-trimethylpentane. These alkylates are produced by the aid of acidic-type catalysts.

Many types of alkylating agents are used, but those that easily form carbonium ions are the most important. Olefins, alcohols and alkyl halides are the most commonly used alkylating agents.

4.1. Alkylation Catalysts

Catalysts for alkylation reactions used essentially to be of the homogeneous type. However, due to the discovery that certain molecular sieves (zeolites) are not only acidic enough to promote carbonium ion or ion-pair formation, but are also more selective than homogeneous catalysts, they are now being used in some new alkylation processes.

Homogeneous alkylation catalysts may be Brønsted acids such as HF and H_2SO_4 etc. Using alkenes as alkylating agents, a proton is donated by the acid to the substrate.

$$>C=C< + H^+ \rightleftharpoons -\overset{\overset{H}{|}}{\underset{|}{C}}-\underset{|}{C^+}-$$

When alkenes are treated with a Lewis acid such as $AlCl_3$, a small amount of a proton acid is normally added as a co-catalyst to promote the formation of carbonium ions.

$$>C=C< + HCl + AlCl_3 \rightleftharpoons -\overset{\overset{H}{|}}{\underset{|}{C}}-\underset{|}{C^+}- + AlCl_4^-$$

If alcohols are the alkylating agents in the presence of Brønsted acids, they are protonated and carbonium ions may be formed.

$$ROH + H^+ \rightleftharpoons \left[\underset{H}{R\overset{+}{O}H}\right] \rightleftharpoons R^+ + H_2O$$

In presence of Lewis acids, such as $AlCl_3$, a complex is first formed with the alcohol and HCl is released. The complex then donates the carbonium ion.

$$ROH + AlCl_3 \rightarrow ROAlCl_2 + HCl$$

$$ROAlCl_2 \rightleftharpoons R^+ + {}^-OAlCl_2$$

Alkyl halides have been extensively used to alkylate aromatic compounds. When $AlCl_3$ is used as the catalyst, the reaction is normally referred to as Friedel-

Crafts alkylation. The first step in alkylating aromatics, is the formation of a carbonium ion.

$$RCl + AlCl_3 \rightarrow R^+ + AlCl_4^-$$

The aromatic ring is then attacked by the carbonium ion which substitutes a proton from the ring. One drawback of the liquid-phase Friedel-Crafts type reactions is the highly corrosive nature of the catalyst, which requires specially constructed materials for the reactor. Additionally, the alkylated product needs further treatment via caustic and water washing to remove traces of $AlCl_3$ or BF_3.

Alkylation with heterogeneous catalysts has been carried out using a variety of acidic oxides such as Al_2O_3 and Al_2O_3/SiO_2. These catalysts also promote carbonium ion-type reactions. Depending on the method of preparation, silica/alumina catalysts may be amorphous or crystalline. These compounds have both Brønsted and Lewis acid sites. When alkylating benzene with ethylene, on a zeolite catalyst for example, adsorbed ethylene is protonated at a Brønsted acid site on the catalyst surface forming an ethylcarbonium ion.

$$Zeol{-}O^-H^+ + CH_2{=}CH_2 \rightarrow CH_3{-}CH_2^+ + Zeol{-}O^-$$

The carbonium ion then attacks the benzene ring giving ethylbenzene and the proton is regained by the zeolite.

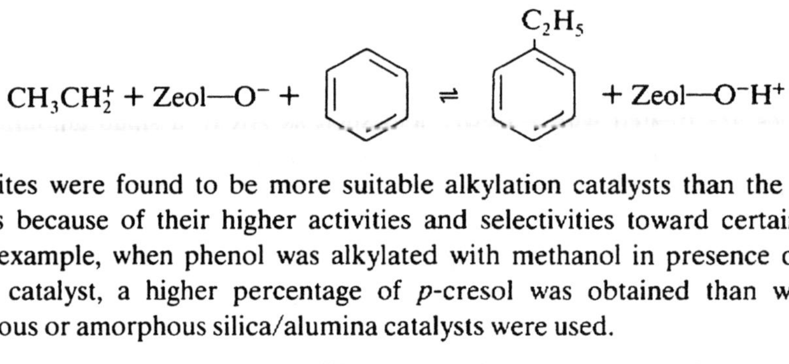

Zeolites were found to be more suitable alkylation catalysts than the amorphous types because of their higher activities and selectivities toward certain reactions. For example, when phenol was alkylated with methanol in presence of a zeolite-type catalyst, a higher percentage of *p*-cresol was obtained than when homogeneous or amorphous silica/alumina catalysts were used.

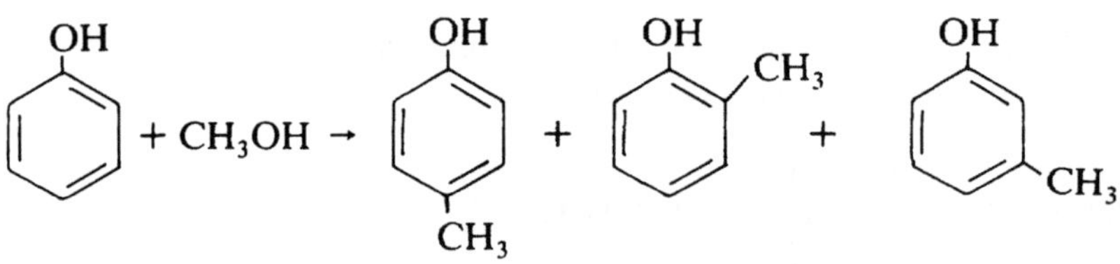

This was explained on the basis of the shape selectivity and the balanced acidity of these zeolites. A short review of zeolite structure is important for explaining these properties.

Zeolites

Synthetic zeolites are crystalline compounds which have similar structure to natural faujasite and mordenite. Zeolites can be synthesized by reacting sodium aluminate and sodium silicate in presence of sodium hydroxide under proper reaction conditions.

$$NaAl(OH)_4(aq) + Na_2SiO_3(aq) + NaOH(aq) \rightarrow Na_x[(AlO_2)_x(SiO_2)_y]\, z\, H_2O$$

The basic building unit in zeolites is a tetrahedron made of four oxygen anions which surrounds either a silicon or an aluminum cation. Each oxygen anion (−2 oxidation state) is shared between either two silicon, two aluminum or an aluminum and a silicon cation. The four oxygen anions of the silicon tetrahedron (SiO_4) are balanced by the +4 oxidation state of the silicon cation, while the four anions of the aluminum tetrahedron (AlO_4) are not balanced, and have −1 net charge. Excess charges of the alumina tetrahedra are balanced by protons or metal cations such as Na^+, K^+, Ca^{2+}, Mg^{2+}, La^{2+} etc. A general formula for zeolites could be represented as:

$$M_{x/n}[(AlO_2)_x(SiO_2)_y]\, z\, H_2O$$

where M represents the metal cation, n the oxidation number of the cation, and x, y and z are integers.

A number of zeolites have been synthesized, but only few have catalytic activity. Among the important zeolites used as catalysts are X- and Y-zeolite, ZSM-5 and ZSM-34.

Zeolites differ in their crystalline structure, the Al/Si ratio, the cation type and its location within the zeolite structure. In forming zeolites, the silica and the alumina tetrahedra join in a special arrangement forming a framework structure that encloses a cavity. In X- and Y-zeolites, for example, a cuboctahedron (sodalite unit) is formed from 24 silica and alumina tetrahedra. See Fig. 4-1.

Sodalite units join together forming a unit cell. A structure made from 10 sodalite units incorporates the cavity or the supercage. The opening of the pores to the supercage is surrounded with 6 sodalite units forming a 12-oxygen-ring structure and it has a diameter of 7.4 Å (Fig. 4-2). Pore dimensions of important zeolites are shown in Table 4-1. As can be seen from the table, zeolites having relatively small, medium and large pores, can be synthesized.

Some zeolites are highly selective catalysts, and this has been correlated with the ability of the reacting molecules to diffuse into, and the ability of the product

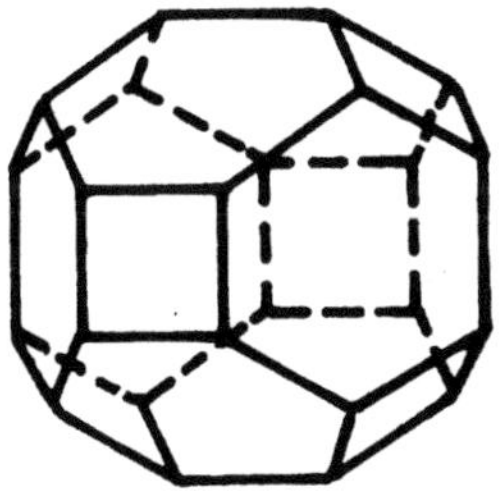

Fig. 4-1. Representation of a Sodalite Unit made from 24 silica and alumina tetrahedra. (Intersecting points represent silicon or aluminum while the lines represent oxygen anions)

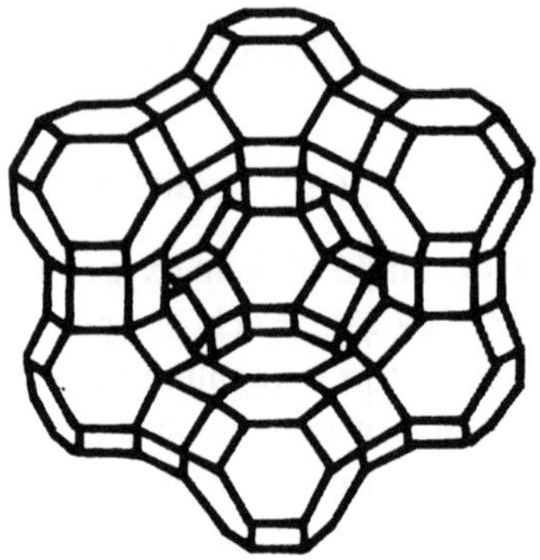

Fig. 4-2. A representation of a zeolite pore made of 12-oxygen ring structure.

Table 4-1. Pore dimensions of some zeolites.

Zeolite Type	Tetrahedral Units	Dimensions Å
X-, Y-Zeolite	12	7.4
ZSM-4	12	7.4
Mordenite	12	6.7 × 7.0
Offretite	12	6.4
ZSM-5	10	5.4 × 5.6
Zeolite A	8	4.1

molecules to diffuse out of the supercage. The supercage is considered the locus of the catalytic activity of zeolite catalysts. If a molecule is too large, it cannot diffuse into the supercage and accordingly would not react. If, on the other hand, a large molecule is formed in the supercage it cannot diffuse out. For example, when ZSM-5 (an intermediate pore zeolite) was used for alkylating benzene with ethylene, less polyalkylated benzenes were obtained than from alkylations with $AlCl_3$ or with zeolites with larger pores. When mordenite (pore diameter 6.7 Å) was used to catalyze the alkylation of benzene with methanol, the main product was a mixture of polyalkylated benzenes with a carbon number of 11 or higher.

An important property relating to the activity of zeolites is their acidity. Zeolites, like silica-alumina catalysts, have both Brønsted and Lewis acid sites. Brønsted acid sites (strong acid sites) are caused by the silanol groups (SiOH) in the zeolite surface. In one type such as HY-zeolite, protons can bond with oxygen anions giving the —OH of the silanol. This could simply be represented as:

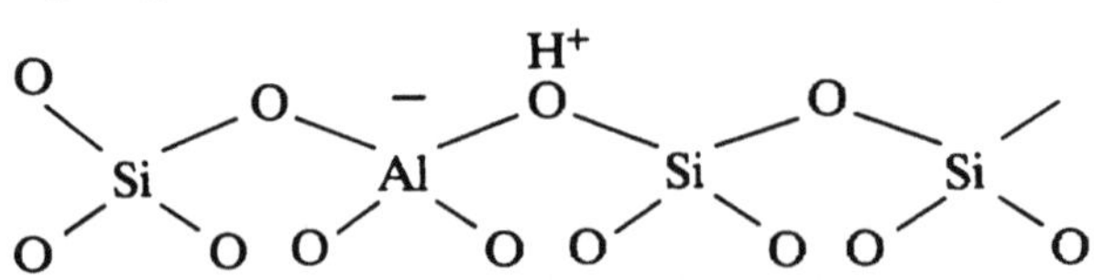

When HY-zeolite is heated to temperatures in the range of 400—500°C, dehydroxylation takes place to a structure having Lewis acid sites. It seems that for each Lewis acid site produced, two Brønsted acid sites are lost.

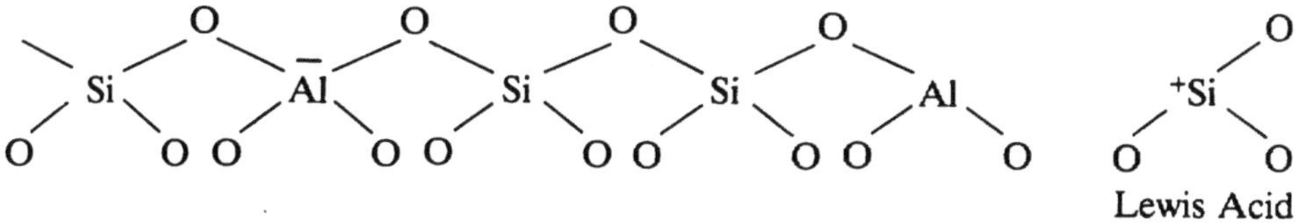

The higher activity of zeolites compared to silica-alumina catalysts may be attributed to a greater number of active acid sites and to their higher adsorption power for the reactant molecules due to their fine structure. Infrared spectroscopy has been used to identify the —OH of the silanol groups. Investigating H—ZSM—5 has shown two absorption bands at 3720 and 3605 cm^{-1}. These were found to correspond to the presence of —OH groups which terminate the exterior faces of the zeolite crystal, the 3605 cm^{-1} absorption being that of the more acidic group. These silanol groups interact with weak bases such as pyridine and ammonia. Studies of pyridine or ammonia adsorption—desorption characteristics are important in revealing strong acid sites of the zeolite surface.

Another factor that affects the catalytic activity of zeolites is the type of cations present. For example, NaY-zeolite was found to be almost inactive in alkylating toluene with methanol. When Na^+ was exchanged with the divalent cation Ca^{2+}, the alkylation was more effective. This may be a manifestation of the fact that Ca^{2+} ions have the ability to split off the adsorbed water molecules and to release protons. Protons that are released from the adsorbed water become associated with oxygen anions in the zeolite lattice and show IR absorption bands at 3650 and 3450 cm^{-1}. These protons are responsible for the stronger acidity of CaY- over NaY-zeolite.

4.2. Alkylation Using Alkenes (Olefins)

Akenes have been extensively used as alkylating agents because of their reactivity toward the formation of carbonium ion intermediates. As electrophiles, carbonium ions can attack a negative center such as an alkene, an aromatic nucleus or even a hydrocarbon. In the latter case the hydrocarbon has to have a labile hydrogen that can leave as a hydride ion, yielding a stable carbonium ion.

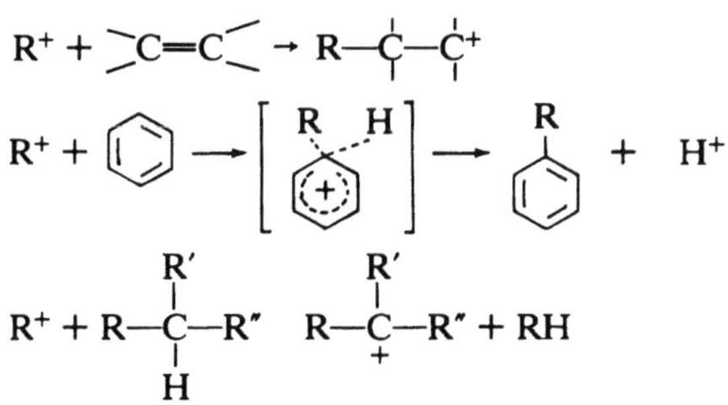

a stable carbonium ion

4.2.1. ALKYLATION OF ISOBUTANE (PRODUCTION OF 2,2,4-TRIMETHYLPENTANE AND OTHER BRANCHED ALKANES)

Alkylation of isobutane with light olefins produces a highly branched alkylate. The product has a high octane rating and is used as a blend with gasoline to raise its octane number. Alkylation of isobutane with isobutene produces 2,2,4-trimethylpentane. This compound is used as a standard for the assessment of the octane number of different gasolines. A research octane number (RON) of 100 has been assigned to this compound in comparison to a zero RON for *n*-heptane. By preparing a blend from the two compounds in different proportions, the octane number of a gasoline sample may be determined. Other important properties of alkylates are; low vapor pressure, high heat of combustion per unit mass, and a good boiling range.

Catalysts normally used for the alkylation are sulfuric acid and anhydrous hydrofluoric acid. The first step in the alkylation is the addition of a proton to the olefin, forming a carbonium ion. When isobutene is used as the alkylating agent *t*-butyl carbonium ion is formed.

$$CH_2{=}C(CH_3){-}CH_3 + H^+ \rightarrow (CH_3)_3C^+$$

Tertiary butyl carbonium ion then adds to another isobutene molecule forming a new carbonium ion.

$$(CH_3)_3C^+ + CH_2{=}C(CH_3){-}CH_3 \rightarrow (CH_3)_3C{-}CH_2{-}C^+(CH_3)_2$$

The new carbonium ion abstracts a hydride ion from isobutane giving 2,2,4-trimethylpentane and *t*-butyl carbonium ion.

$$(CH_3)_3C{-}CH_2{-}C^+(CH_3)_2 + (CH_3)_3CH$$
$$\rightarrow (CH_3)_3C{-}CH_2{-}CH(CH_3)_2 + (CH_3)_3C^+$$

The newly formed carbonium ion from the last step adds to an isobutene molecule until all the available isobutene is consumed.

Other alkylates are produced by reacting isobutane with different light olefins. For example, when propene is used as the alkylating agent, a mixture of 2,2-dimethylpentane (60—80%), 2-methylhexane (10—30%) and 2,2,3-trimethylbutane (7—11%) is produced.

The use of 1-butene, 2-butene or a mixture of both to alkylate isobutane, produces an alkylate of a high octane quality. Either HF or H_2SO_4 may be used to catalyze the reactions. Similar to isobutane alkylation with isobutene and propene, reactions of *n*-butenes take place via a carbonium ion intermediate. In case of 1-butene, the main product is 2,2-dimethylhexane while, in 2-butene, the dominant product is 2,2,3-trimethylpentane. For 1-butene the sequence of reactions could be represented as:

$$CH_3{-}CH_2{-}CH{=}CH_2 + H^+ \rightarrow CH_3{-}CH_2{-}CH^+{-}CH_3$$

$$CH_3CH_2CH^+{-}CH_3 + CH_3{-}CH(CH_3){-}CH_3 \rightarrow CH_3CH_2CH_2CH_3 + CH_3{-}\overset{+}{C}(CH_3){-}CH_3$$

$$CH_3{-}\overset{+}{C}(CH_3){-}CH_3 + CH_3{-}CH_2{-}CH{=}CH_2 \rightarrow (CH_3)_2C(CH_3){-}CH_2{-}C^+H{-}CH_2{-}CH_3$$

$$(CH_3)_2C(CH_3){-}CH_2C^+HCH_2CH_3 + CH_3CH(CH_3)CH_3 \rightarrow (CH_3)_2C(CH_3){-}(CH_2)_3CH_3 + CH_3\overset{+}{C}(CH_3)CH_3$$

The *t*-butyl carbonium ion produced in the last step continues another chain reaction. A similar reaction sequence takes place with 2-butene. Properties of alkylates from 1-butene and 2-butene are shown in Table 4-2.

Isobutane alkylation with olefins is normally run at low temperatures (approximately 60°C). A sufficient pressure is used to keep the reactants in the liquid

Table 4-2. Properties of isobutane alkylates from 1- and 2-butenes catalysed with H_2SO_4.

	1-Butene	2-Butene
H_2SO_4 conc. wt %	99.7	99.5
Alkylate properties		
RON clear	99.6	99.8
MON clear	94.5	94.8
ASTM, end point °F	303	301
C_8 compounds %	89.88	89.32
C_9^+ compounds %	4.76	4.13

Table 4-3. Ranges of operating conditions for H_2SO_4 and HF catalyzed alkylations.

Process conditions	H_2SO_4	HF
Temperature °C	2–16	16–52
Isobutane/Olefin	3–12	3–12
Contact time, Min	20–30	8–20
Acid strength wt %	88–95	80–95
Acid in emulsion vol. %	40–60	25–80

phase. A high isobutane to olefin ratio improves the yield as well as the octane number of the product alkylate. Alkylation conditions for sulfuric acid and hydrofluoric acid catalyzed reactions are similar to those shown in Table 4-3.

4.2.2. ALKYLATION OF BENZENE WITH ALKENES

Benzene is a reactive aromatic hydrocarbon which is mainly produced from naphtha cat-reforming (see Chapter 3). Under proper reaction conditions, benzene can be substituted with different electrophilic reagents. When alkylated, the benzene ring is activated and di-, tri- and polyalkylated compounds are also produced. Monoalkylated products could be maximized however by increasing benzene/olefin ratio. Benzene may be alkylated with lower olefins such as ethylene and propylene, or with higher olefins (C-12—C-14) producing alkylates for detergent use.

4.2.2.1. *Production of Ethylbenzene*

Ethylbenzene is produced by alkylating benzene with ethylene. A small quantity (about 10%) is obtained by the superfractionation of the C-8 aromatic cut of the reformate and other refinery streams. Both liquid and gas phase alkylations are used. In the liquid phase mode, $AlCl_3$ promoted with HCl or ethylchloride is the catalyst.

The alkylation reaction is exothermic and is conducted at about 150°C and 5—10 bar.

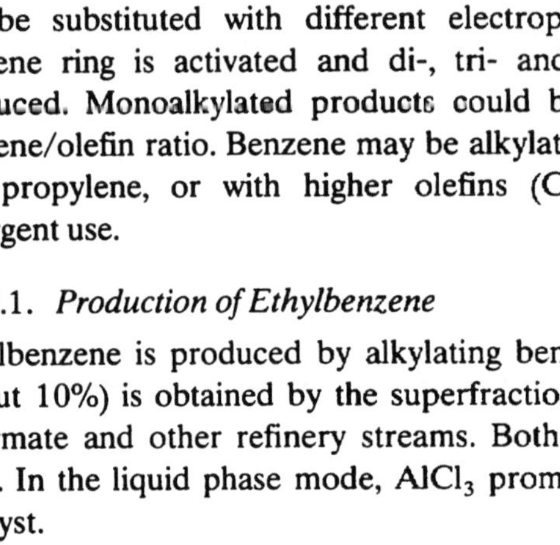

Diethylbenzene and higher alkylates are also formed during the reaction. A higher benzene-to-ethylene ratio decreases the amount of polyalkylates. Recycling higher benzene alkylates increases ethylbenzene yield through transalkylation.

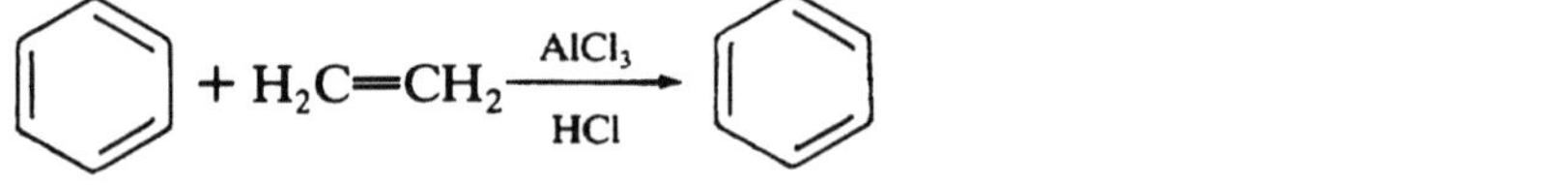

Selectivity to ethylbenzene is also increased by limiting the conversion to about 50%. Newer processes, using a low $AlCl_3$ concentration, a very low ethylene-to-benzene ratio, and a higher temperature (about 150°C), can reach a selectivity of over 99% for ethylbenzene.

Recently, a fixed-bed, solid-catalyst, gas-phase process has been developed by Mobil-Badger. Catalysts used are of the crystalline molecular sieve type (ZSM-5).

In alkylating benzene over zeolite catalysts, secondary reactions may take place such as cracking, (zeolites are also important cracking catalysts) oligomerizations etc. Accordingly, alkyl aromatics not normally found in Friedel-Crafts alkylations are formed.

$$2\ CH_2{=}CH_2 \rightleftharpoons C_4H_8$$

$$C_4H_8 + CH_2{=}CH_2 \rightleftharpoons C_6H_{12}$$

$$C_6H_{12} \xrightarrow{\text{cracking}} 2\ C_3H_6$$

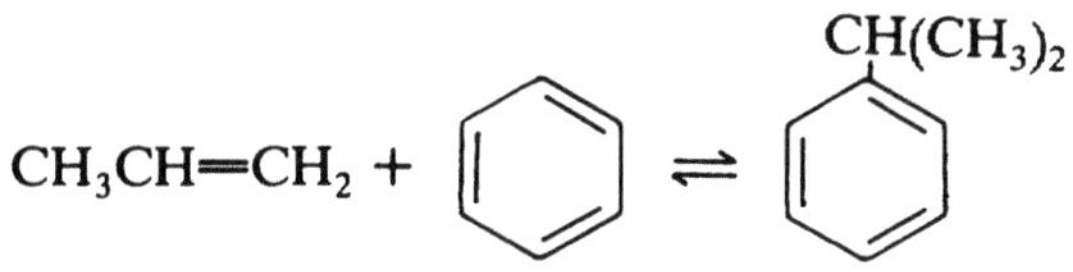

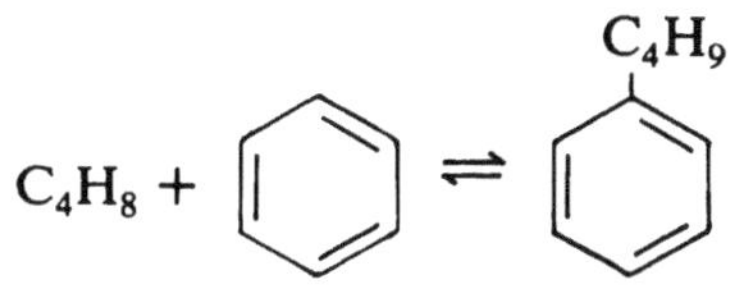

However, excessive oligomerization and cracking have not been observed with ZSM-5 catalysts. This may be due to the wide spacing of the Brønsted acid sites in these catalysts.

The main use of ethylbenzene is the production of styrene by dehydrogenation.

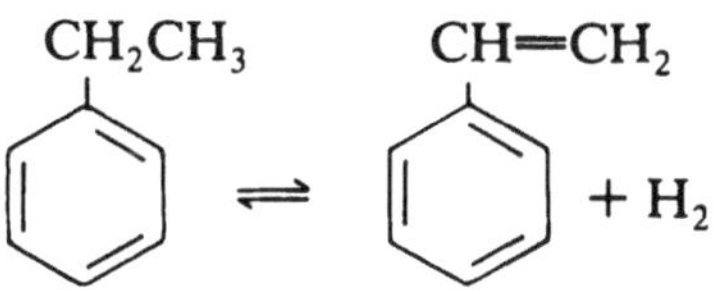

4.2.2.2. *Production of Cumene*

Cumene (isopropylbenzene), is produced by alkylating benzene with propene.

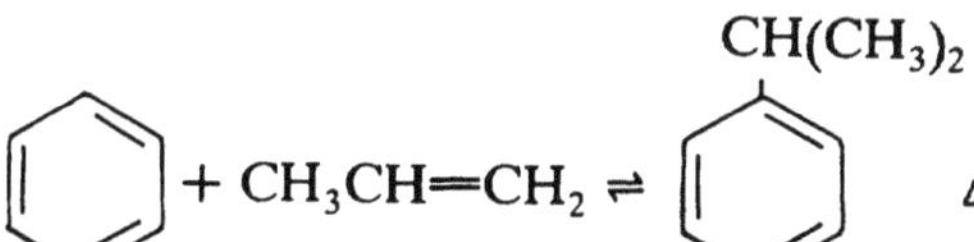

$\Delta H = -113$ kJ/mol

The significance of this process is that it is the only method for the production of cumene. Either a liquid or a gas phase process is used for the alkylation reaction. In the liquid-phase process, the catalyst may be a Lewis acid or a proton acid. The reaction takes place at a relatively low temperature of about 50°C and a pressure of about 5 bar. To reduce polyalkylation, a mixture of propene/propane is used. The presence of small amounts of ethylene can be tolerated since, under propylene alkylation conditions, ethylene is quite unreactive. Polyalkylation can also be reduced by increasing the benzene/propylene ratio. Presence of some butylenes does not represent a serious problem since butylbenzenes can be removed as bottoms from the cumene column.

The main by-products are *n*-propylbenzene, di-, and tri-isopropylbenzene which may be easily separated from cumene by distillation.

In the gas-phase process a SiO_2/H_3PO_4 or H_3PO_4 on Kieselguhr is used as a catalyst at a temperature of about 250°C and a pressure of about 40 bar. A selectivity of about 97% based on benzene can be reached.

The main use of cumene is for phenol production. An important co-product is acetone.

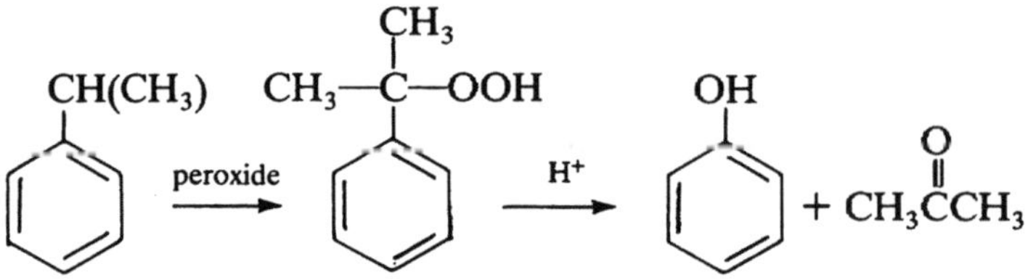

The cumene process is quite significant since it accounts for over 90% of phenol production in the Western World. It is also one of the most competitive processes for acetone production.

Phenol is a precursor of many important chemicals and polymers. Phenolic resins are one of the oldest thermosetting plastics with good chemical properties. When reacted with acetone, bisphenol A is produced, which is an important monomer for polyurethanes, epoxy resins and polycarbonates.

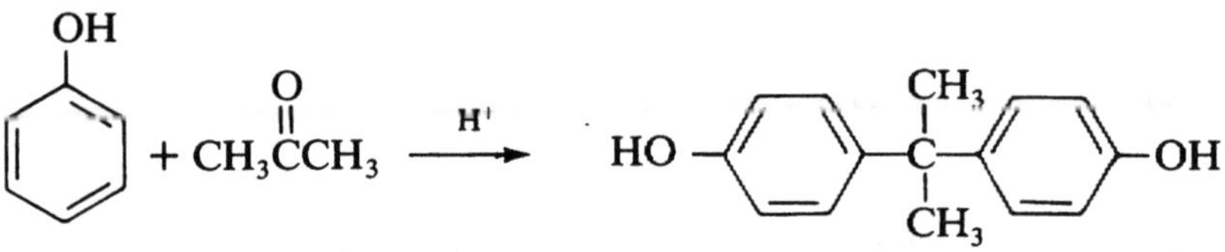

Phenol is also used for producing salicylic acid, acetylsalicylate and 2,4-dichlorophenoxyacetic acid (2,4-D) which is a selective weed killer.

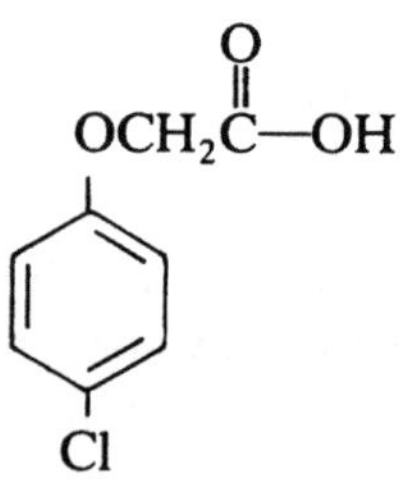

4.2.2.3. *Production of Benzene-detergent Alkylates*

Benzene alkylates were originally produced by alkylating benzene with tetrapropylene (propylene tetramer). Detergents produced from these alkylates were nonbiodegradable due to the branching of the side chain.

Alkylation of benzene with linear C-10—C-14 alkenes (linear monoolefins) produces a mixture of linear alkylbenzenes (LAB). Sulfonation followed by neutralization produces linear alkylbenzene sulfonates (LAS) which are biodegradable detergents. Using α-olefins for the alkylation the reactions could be represented as:

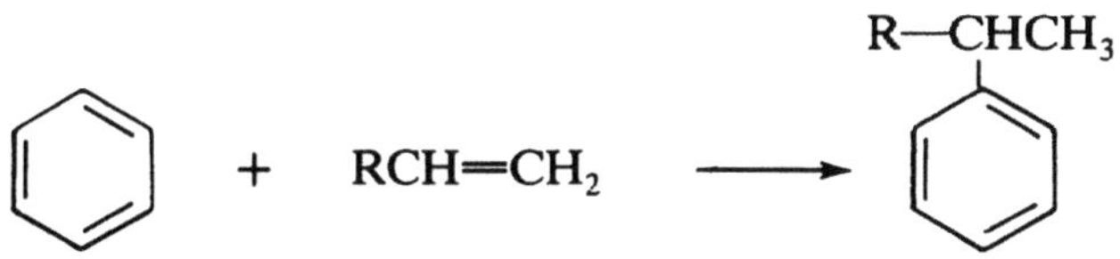

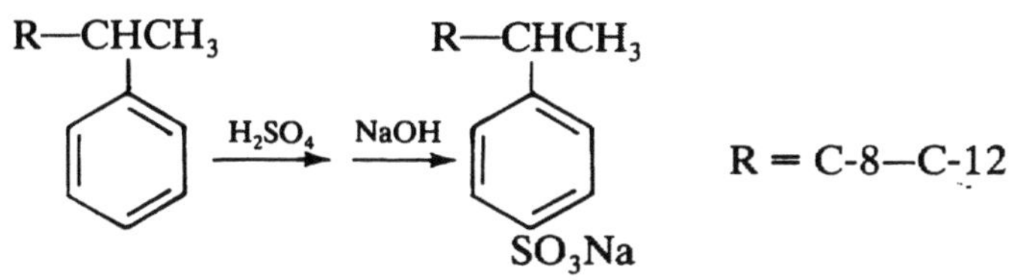

Linear alkenes used for the alkylation of benzene are produced by dehydrogenation of n-alkanes extracted from kerosine fractions, wax cracking, dehydrochlorination of monochlorinated n-alkanes or from ethylene oligomerization using Ziegler catalysts. The latter process produces α-olefins. The alkylation may be affected in a liquid phase process using HF at a temperature of about 70°C. Other catalysts such as HBF_4 or $AlCl_3$ may also be used. In the presence of proton acids,

Table 4-4. Typical properties of linear alkylbenzene (LAB) for detergent use.

Test	Result
Bromine number	0.02
Alkylbenzene content wt %	97.4
Doctor test	Negative
Unsulfonatable content wt %	1.0
Water wt %	0.1
Specific gravity @ 60 °F	0.8612
Refractive index $n_{D^{20}}$	1.4837
Average molecular weight	240
Distillation (ASTM D-86) °F	
IBP	538
10 vol. %	547
20 vol. %	550
50 vol. %	554
70 vol. %	559
90 vol. %	569
95 vol. %	576
EP	589
Biodegradability (ASTM D-2667) %	95.0

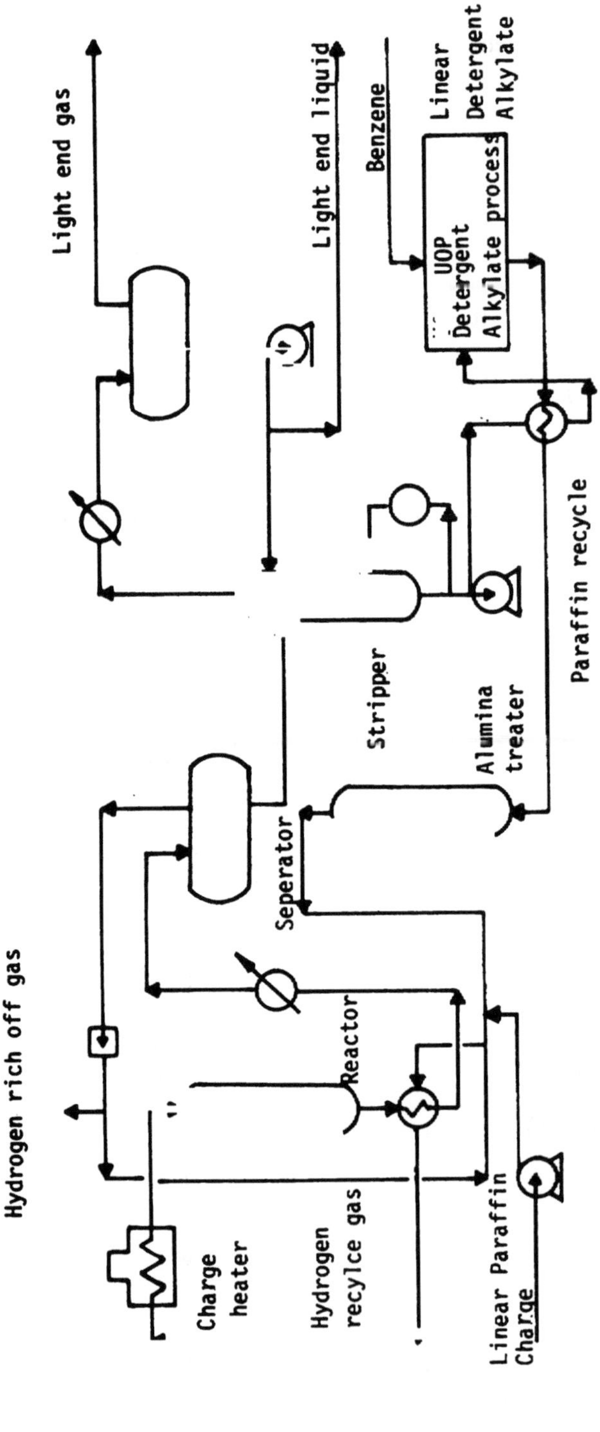

Fig. 4-3. Pacol Process for linear alkyl benzene production from n-paraffins.

a carbonium ion is formed and double bond isomerization may take place. Depending on the position of the double bond in the alkylating olefin, benzene could be situated at any position along the chain of the attacking olefin.

In the gas phase process (UOP), the alkylation is integrated with an *n*-alkane dehydrogenation. Monoolefins (C-10—C-15) with some *n*-paraffins from the dehydrogenation section are fed with benzene to the alkylation section in the presence of HF. Excess benzene, unreacted *n*-alkanes, HF catalyst, linear alkylbenzene and heavy alkylate are separated by sequential fractionation. Fig. 4-3 shows the Pacol process for producing linear alkylbenzene from *n*-alkanes. Linear alkylbenzene produced by this method has a low olefin content and a relatively narrow distillation range. Typical benzene alkylate (for detergents) properties are shown in Table 4-4. Alkylbenzene sulfonates are the most important anionic detergents and they account for approximately 50% of the total demand for anionic detergents.

4.2.3. ALKYLATION OF PHENOL WITH ALKENES

Alkylation of phenol with alkenes is similar to benzene alkylation. Alkenes in the range of C-8—C-12 are normally used to produce alkylphenols. The alkylation is conducted in the liquid phase at about 50°C, using HF or BF_3 as catalysts. Phenol alkylates are normally ethoxylated with ethylene oxide and the products are used as nonionic detergents. Alkylphenols could also be sulfonated to anionic detergents.

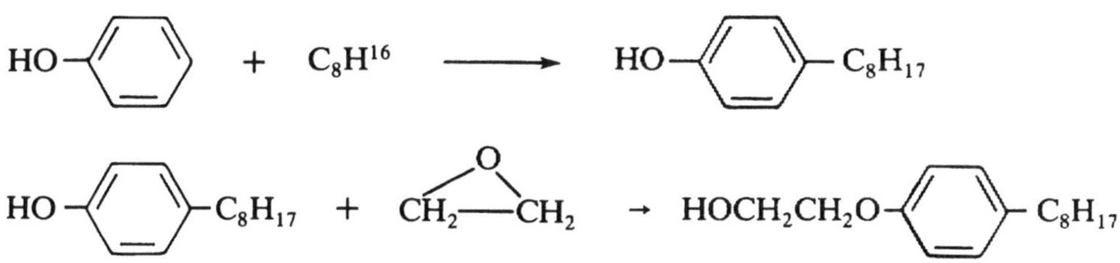

However, their importance is gradually decreasing because of their nonbiodegradable nature.

4.3. Alkylations Using Alcohols

As alkylating agents, alcohols are not used as much as alkenes probably because they are less reactive in forming the carbonium ion intermediate. Also, alcohols normally used for alkylations are produced from alkenes. As mentioned earlier, catalysts used for alkylations with alcohols are either Brønsted or Lewis acids. Heterogeneous catalysts have recently been tried to produce some important alkylates. For example, ethyl alcohol has been tried in the alkylation of benzene for ethylbenzene production. Methyl alcohol has also been used to alkylate toluene for the selective production of *p*-xylene.

4.3.1. ALKYLATION OF PHENOL WITH METHANOL (PRODUCTION OF METHYLPHENOLS)

The alkylation of phenol with methyl alcohol is a commercially viable process for producing cresols and xylenols.

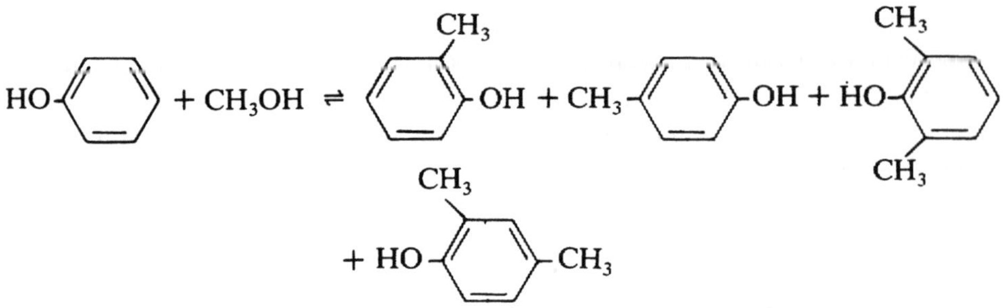

The reaction is performed in the liquid phase at about 300°C and 50 bar using an aluminum oxide catalyst.

The ratio of dimethylated to monomethylated phenols depends on the reaction conditions and on the methanol/phenol ratio. A high methanol/phenol ratio increases the yield of dimethylphenols. Lower temperatures and pressures generally favor the formation of monomethylated phenols and specially *o*-cresol. Higher temperatures and pressures favor the formation of 2,6- over 2,4-xylenol.

Recent work with HY-, HKY-zeolites and ZSM-5 has shown that phenol alkylated with methanol gave a higher percentage of *p*-cresol due to the shape selectivity of these catalysts.

Cresols in general are not used in the acid form. A cresol mixture is utilized for the production of tricresylphosphates which are good plasticizers for vinyl chloride polymers and are important gasoline additives. A mixture of *o*- and *m*-cresol is used for the production of phenol-formaldehyde resins. 2,6-Xylenol is used for the production of polyphenylene oxide, a polyether with good heat and chemical resistance. *o*-Cresol is mainly used for the production of herbicides and insecticides.

4.3.2. ALKYLATION OF TOLUENE WITH METHANOL (PRODUCTION OF XYLENES)

Xylenes are mainly produced by a distillation-extraction scheme from reformed gasolines. Xylenes are important precursors for many chemicals (see Chapter 1). The ratio of the isomers, *para* : *ortho* : *meta* is approximately 1 : 1 : 2, which is very close to the thermodynamic equilibrium values for the three isomers at the reforming temperature (Table 4-5).

Table 4-5. Thermodynamic equilibrium values for xylene isomers at three temperatures.

Aromatics wt %	Composition 200°C	300°C	500°C
p-Xylene	21.8	21.1	18.9
o-Xylene	20.6	21.6	23.0
m-Xylene	53.5	51.1	47.1
Ethylbenzene	4.1	6.2	11.0

p-Xylene is the most important of the three isomers since it is utilized for the synthesis of terephathalic acid for polyester production. A process that selectively produces *p*-xylene may be economically competetive to the currently used scheme.

Recent work for alkylation of toluene with methanol using HY-zeolite and ZSM-5 at a temperature range of 300—700°C has shown that over 90% selectivity for *p*-xylene could be reached. If this process proves feasible, the alkylation of toluene over shape-selective zeolites may be utilized in the future.

4.3.3. ALKYLATION OF AMINES WITH METHANOL (PRODUCTION OF METHYLAMINES)

Methylamines can be synthesized by alkylating ammonia with methyl halides and methyl alcohol. For large scale production, however, methyl alcohol is used as the alkylating agent. The reaction usually takes place at approximately 500°C and 20 bar using either aluminum silicate or phosphate catalyst. The alkylation does not stop at the monomethylamine stage since the amine produced is a better nucleophile than ammonia. Alkylation continues to di- and trimethylamines. The product distribution at equilibrium (450°C) is: monomethylamine (MMA) 43%, dimethylamine (DMA) 24% and trimethylamine (TMA) 33%.

$$CH_3OH + NH_3 \rightleftharpoons CH_3NH_2 + H_2O$$

$$CH_3OH + CH_3NH_2 \rightleftharpoons (CH_3)_2NH + H_2O$$

$$CH_3OH + (CH_3)_2NH \rightleftharpoons (CH_3)_3N + H_2O$$

Dimethylamine is the most widely used of the three amines and its relative yield can be increased by recycling. Excess ammonia also increases the yield of the mono- and the dimethylamines. For example, increasing the ammonia/methanol ratio to 2.4 at 500°C increases the yield of monomethylamine to about 55% and dimethylamine to about 26%. Separation of the methylamines is done by a distillation-extraction technique.

Monomethylamine is used in the synthesis of Sevin, a widely used insecticide. It is also used for the production of other pesticides and pharmaceuticals.

Dimethylamine is the most important of the three amines since it is used for the synthesis of dimethylformamide and dimethylacetamide (both compounds are solvents for acrylic and polyurethane fibers). It is also used in the synthesis of various surface active agents and rubber chemicals as well as unsymmetrical dimethylhydrazine (a rocket fuel).

Trimethylamine has only one major use, that is the synthesis of choline salts which are high-energy additives for poultry feed.

4.4. Alkylation using Alkyl Halides

Alkyl halides are general alkylating agents for a variety of compounds. They may be used to synthesize aromatic alkylates (Friedel-Crafts), ethers (Williamson synthesis), thioethers and amines. They are used in acetoacetic ester synthesis and in preparing Grignard reagents. However, alkyl halides are not widely used in industrial applications since, in most of their reactions, hydrogen chloride is released, which is corrosive. Special construction materials for the reaction vessels should be used.

In alkylating ammonia, a mixture of amines is produced.

$$NH_3 + RX \rightarrow RNH_2 + HX$$

$$RNH_2 \xrightarrow{RX} R_2NH \xrightarrow{RX} R_3N \xrightarrow{RX} R_4N^+X^-$$

The reaction, however, is not very feasible for preparing primary and secondary amines since they are stronger nucleophiles than ammonia, as mentioned before (Section 4.3.3). The yield of primary and secondary amines could be improved by increasing the NH_3/RX ratio. Alkylation of compounds that possess electron withdrawing groups, however, could be effective since the amines formed are weaker bases than ammonia. For example, α-amino acids could easily be synthesized from α-halo acids and ammonia. The amino acid produced is a weaker base than ammonia due to the inductive effect of the carboxyl group.

$$NH_3 + \underset{\displaystyle Cl}{RCHCOOH} \rightarrow \underset{\displaystyle NH_2}{RCHCOOH} + HCl$$

The industrial method for producing amines using methanol has already been dealt with in Section 4.3.3. However, in certain industrial applications, alkyl halides are used, as in the case of alkylating benzene with monochloroalkanes. In this process, chloroalkanes are obtained by chlorinating a mixture of *n*-alkanes (*n*-paraffins) in the range of C-12—C-15. After separation of the monochloroalkanes, they may either be directly used as alkylating agents for benzene in presence of $AlCl_3$ or dehydrochlorinated to linear alkenes (linear olefins). As mentioned earlier, linear olefins are used to alkylate benzene. In either case the benzene alkylate is processed to linear alkylbenzene sulfonate (LAS) for detergent production.

References

Dyer, F.: *Catalysis of Organic Reactions* (Edited by William Moser), (Mobil/Badger Ethylbenzene Process), Marcell and Dekker Inc., 1984.

March, J.: *Advanced Organic Chemistry: Reactions, Mechanisms and Structure*, McGraw-Hill Book Co., N.Y., 1968.

Matar, S. and Muse, M.: 'Effect of Cation Exchange on Y-Zeolite Activity in Toluene Alkylation with Methanol', *Arabian Journal for Science and Engineering* **12**(1), 1986.

Yashima, T., Sakaguchi, Y., and Namba, S.: 'Selective Formation of *p*-Xylene by Alkylation of Toluene with Methanol on ZSM-5 Type Zeolites' presented at the International Congress of Catalysis, Tokyo, 1980.

Gates, B., Katzer, J., and Schuit, G.: *Chemistry of Catalytic Processes*, McGraw-Hill Book Co., N.Y., 1979.

Venuto, P., Hamilton, L., Landis, P., and Wise, J.: *Journal of Catalysis* (5) 1966.

Olah, G.: *Friedel-Crafts and Related Reactions*, Interscience Publishers, *11*(1), 1964, p. 3.

Frank, B. *et al.*: 'Reforming for BTX' SRI Report # 129, May 1980.

Kaeding, W., Chu, C., Young, L., Weinstein, B., and Butter, S.: *Journal of Catalysis* (67) 1981.

Vora, B., Pujado, P., and Spinner, J.: 'Latest LAB Developments', *Hydrocarbon Processing* **63**, 1984, p. 11.

CHAPTER 5

Catalytic Oxidation Reactions

5. Introduction

Catalytic oxidation of hydrocarbons is the largest category of catalytic organic reactions in the petrochemical industry. If oxidation is considered in the formal way i.e., increase in the formal oxidation number of the carbon atom, then reactions such as dehydrogenation should also be classified as oxidation reactions. In our discussion we shall consider as oxidation reactions all reactions of organic compounds that involve oxygen as a reactant and/or lead to the formation of an oxidized form of the organic reactant. Thus oxidative dehydrogenation and ammoxidation will belong to oxidation reactions. However, we shall be selective in the reactions that we shall cover, for various reasons. The Oxo and dehydrogenation reactions will not be included here since they are covered elsewhere in the book.

Oxidation reactions may be classified into: vapor-phase reactions which are mainly hetergeneously catalysed, and liquid-phase reactions which are commonly homogeneously catalyzed. For some reactions the two categories exist and have both been used industrially with varying importance over time e.g. oxidation of ethylene to acetaldehyde. However, for the purpose of our discussion in this text oxidation reactions will be classified according to the type of organic reactant. We shall therefore be talking about oxidation of saturated hydrocarbons, olefins, aromatic compounds, alcohols and compounds with other functional groups.

5.1. Oxidation of Saturated Hydrocarbons

Synthesis of oxygen-containing compounds by the oxidation of saturated hydrocarbons is not an industrially feasible route. Saturated hydrocarbons are highly stable and their reaction with oxygen is difficult to control once it starts. It tends to oxidize the hydrocarbon all the way to CO_2 and H_2O.

However, there are a few industrial processes based on the oxidation of saturated hydrocarbons. Most significant among these is the steam reforming process of methane (and higher hydrocarbons) to produce a mixture of hydrogen, carbon monoxide and carbon dioxide. This process will be discussed in detail in the chapter on synthesis gas (Chapter 6).

Another process is the controlled oxidation of methane to formaldehyde. This process has been superseded by the oxidation of methanol. The methane oxidation process accounts for only 8% of the current production of formaldehyde. This is contributed by old plants that are on their way to retirement.

A third process is the synthesis of maleic anhydride by the oxidation of butane.

But this process has also been surpassed by using the more reactive benzene as a substrate, as we shall see later.

5.2. Oxidation of Olefins

The most important substrates that are used in industrial catalytic oxidation are olefins, aromatic compounds and alcohols. We shall discuss oxidation of the most common olefins: ethylene and propylene. The oxidation of higher olefins is gaining interest at the present. A process has been developed for the oxidation of isobutene to methacrylic acid. However, it involves hydration of isobutene to *t*-butanol which is then oxidized at 360°C to methacrolein in a fixed-bed reactor containing a multicomponent mixed oxide catalyst. Methacrolein is then oxidized in a second stage to methacrylic acid.

5.2.1. ETHYLENE OXIDATION

The products of the oxidation of ethylene depend on the reaction conditions including the choice of the catalytic system. The most important industrial catalytic oxidation products directly obtainable from ethylene are ethylene oxide and acetaldehyde. In addition to these we shall discuss the manufacture of vinyl acetate from ethylene since it is carried out in an oxidizing atmosphere and is strongly related to the manufacture of acetaldehyde from ethylene.

(a) *Ethylene Oxide*

Ethylene oxide is used relatively little by itself, but it is an important intermediate, especially in the manufacture of ethylene glycol, which accounts for about two thirds of the consumption of ethylene oxide. The world-wide 1980 production of ethylene oxide was in excess of six million metric tons. Production today is almost entirely by direct oxidation of ethylene with air or oxygen over a supported silver catalyst according to the equation

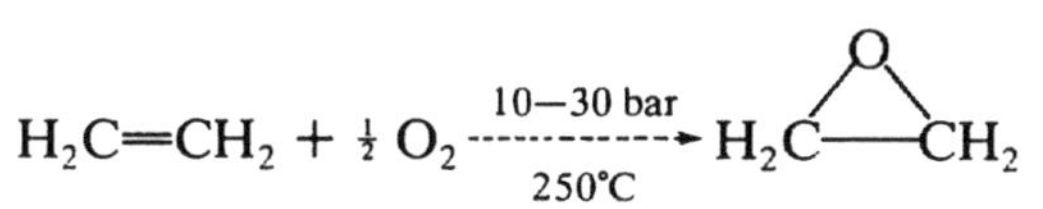

The technology of the process has not changed much since its initial commercialization. Silver is unique in catalyzing the expoxidation of ethylene; all other catalysts are relatively or totally ineffective and the reaction is limited, among lower hydrocarbons, to ethylene. Propylene or butylene do not form epoxides, but styrene, which has no β-hydrogen, is reported to give good yields of styrene oxide.

Side-reactions are the oxidation of both ethylene oxide and ethylene, especially the latter, to carbon dioxide and water with the liberation of large amounts of heat. Much of the effort in catalyst and process development has been directed toward minimizing these side-reactions which cause a yield loss *per se*, as well as catalyst deterioration through excessive temperature rises.

Catalysts are 10—15% silver supported on a low surface area ($<1\ m^2g^{-1}$) alpha alumina. However a fairly large silver area, say 0.05—5.0 m^2/g Ag, is desirable. The catalyst may contain small amounts of a metal promotor such as calcium or barium which acts mainly to stabilize silver against sintering. Overoxidation is minimized by chloride inhibition, the chloride being fed continuously as a few parts per million of organic chlorides, such as 1,2-ethylene dichloride. The role of modifiers has been explained in terms of electronic interactions between silver crystallites and the support, which alter the work function of electrons at the metal surface. Since atomic oxygen causes overoxidation while molecular oxygen enhances ethylene oxide formation, chlorine modifiers are thought to change not only the relative concentration of atomic and molecular oxygen, but also to increase the probability that molecular oxygen reacts to give ethylene oxide.

Early plants used air as the oxidant, but later, despite the added cost, pure oxygen was often used, as it simplifies recycle, allows higher reactor productivity and lowers overall cost. The choice often depends on plant size and local circumstances. When oxygen is used, an inert gas, such as methane, is recycled, which decreases ethylene losses, and decreases the flammability limits.

The highly exothermic oxidation is carried out in a tubular bundle fixed-bed reactor with external cooling. Bundles in excess of 10,000 tubes are normally used. Feed streams contain 20—30% v/v ethylene, 6—8% oxygen, and the remainder being inerts (methane), at 250—330°C and 10—30 bar. Ethylene conversions are limited to about 10% and selectivities to ethylene oxide are 65—75%.

Some theorists project 85.6% selectivity as a theoretical maximum, based on mechanistic considerations. This high selectivity is seldom reached. However, selectivities in the mid 90's have been demonstrated but only at low conversions, 3—4%. The value of 85.6% which is numerically equal to 6/7, is derived from the proposal that ethylene is not adsorbed on silver, but on nondissociatively adsorbed oxygen, which proceeds to give ethylene oxide and an adsorbed oxygen atom.

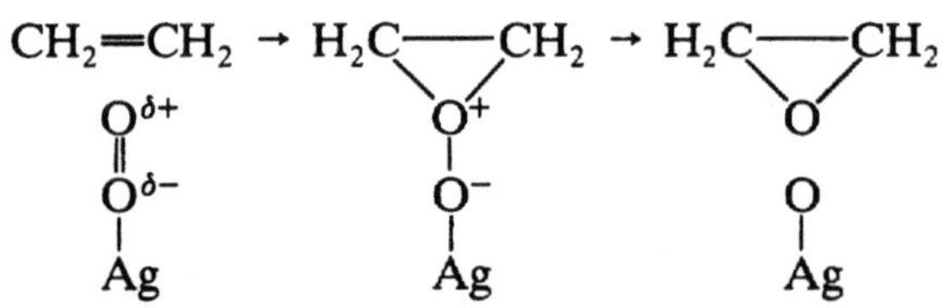

The adsorbed oxygen atoms thus produced react with ethylene to give carbon dioxide and water.

$$6\ O\ (ads) + CH_2{=}CH_2 \rightarrow 2\ CO_2 + 2\ H_2O$$

Overall, of every seven ethylene molecules, six go to ethylene oxide and one to carbon dioxide.

$$7\ C_2H_4 + 6\ O_2\ (ads) \rightarrow 6\ C_2H_4O + 2\ CO_2 + 2\ H_2O$$

The above oxygen adsorption process is assumed to occur when no more ensembles of four virgin silver atoms are available. These ensembles are consumed in a dissociative adsorption of oxygen. This mode of adsorption could be completely inhibited by introduction of chlorine atoms equal to one fourth the number of surface silver atoms. Each further chlorine atom introduced above this amount reduced by two the number of oxygen atoms subsequently adsorbed.

The dissociative mode of adsorption of oxygen is thought to lead to production of carbon dioxide. Chlorine thus preferentially inhibits an unwanted mode of oxygen adsorption. However, too much moderator may destroy the activity of the catalyst. It has been suggested that traces of chlorine atoms present as impurities in the silver catalyst behave as moderators. Thus different extents of purity of the catalyst may lead to different results. This explains the apparently inconsistent results reported in the literature.

(b) *Acetaldehyde*

Acetaldehyde is predominantly made by the direct oxidation of ethylene by the Wacker process whereby an aqueous solution of cupric chloride containing traces of palladium chloride is used as the oxidizing system. The process may operate as a single-stage or in two stages. In the reaction palladium chloride is first reduced to elemental palladium while oxidizing the ethylene to acetaldehyde and is promptly reoxidized to palladium chloride by the cupric chloride. During catalyst regeneration, the cuprous chloride is reoxidized to cupric chloride by oxygen, the ultimate oxidizing agent.

(i) *Single-Step Process.* Ethylene, oxygen, and recycled gas are fed into a vertical reactor where intimate contact is made with the catalyst solution (containing 200 g/l of $CuCl_2$ and 4 g/l $PdCl_2$) under a slight pressure (3 bar). The heat of the exothermic reaction has to be carried away, and this is done by boiling off some of the water. The gases are scrubbed, and the resulting acetaldehyde solution is separated by fractional distillation using the heat of the reaction. The conditions of the reaction are adjusted so that about 40% of the ethylene and all the oxygen are consumed. The unreacted gas is recycled to the reactor apart from a small bleed stream to eliminate impurities in the feed. In this process, practically pure ethylene and pure oxygen are used.

(ii) *Two-Step Process.* In this operation ethylene and air are adsorbed by contact with the solution of catalysts, with the net result that the copper is reduced from cupric to cuprous and the ethylene is converted to acetaldehyde. This operation is carried out under higher pressure (10 bar) of ethylene and air, which are in such proportion as to make complete conversion of ethylene possible. The solution is withdrawn from the reactor. The pressure is reduced and the acetaldehyde flashed off with some of the water of the solution. Acetaldehyde is recovered by fractional distillation. The spent catalyst is taken to a second reactor where it is mixed with air; oxygen reoxidizes the cuprous chloride to cupric and the catalyst solution is recycled to the reactor.

In the two-step process, it is feasible to use hydrocarbon streams that are not pure ethylene, e.g., mixtures of petroleum refinery gases, and it is also possible to use air instead of the pure oxygen required in the single-step process. For these reasons the two-step process has a certain attractiveness to it.

The yield of acetaldehyde is about 95% in either process.

The introduction of the Wacker process constituted a breakthrough in the manufacture of acetaldehyde. It is also an example of the superiority of homogeneous catalysis when the associated engineering problems are surmounted. Table 5-1 summarizes the technical aspects of the Wacker process in comparison with other processes known for manufacturing acetaldehyde.

Table 5-1. Comparison of the Wacker with other known acetaldehyde manufacturing processes.

Starting Materials	Process	Catalyst, Temperature	Yield %	Conversion per pass, %
C_2H_5OH	Vapor phase Heterogeneous catalytic oxidation.	Ag gauze, 300—575°C.	~85—95	~25—35
C_2H_5OH	Vapor phase Heterogeneous catalytic dehydrogenation.	Cr/activated Cu, 260—290°C.	~85—90	~50
C_2 to C_6 saturated hydrocarbons (LPG)	Vapor phase noncatalytic Heterogeneous oxidation.	None, 100—450°C.	~30—40	~25
$CH{\equiv}CH$	Liquid phase Homogeneous catalytic hydration.	18—25% aqueous H_2SO_4 containing $HgSO_4 + Fe_2(SO_4)_3$ 70—100°C.	~95	~50—60
$CH_2{=}CH_2$	Liquid phase Homogeneous 'catalytic' oxidation. (Wacker)	$PdCl_2/CuCl_2$; 100°C.	~95	~95—99

Mechanistic Studies on the Oxidation of Ethylene to Acetaldehyde

The ethylene-to-acetaldehyde Wacker process lends itself to feasible chemical investigation and it has provided a wealth of chemical information. This is mainly because the system is homogeneous and because it is not catalystic in the proper sense. It consists of a series of stoichiometric reactions which when added together give the catalytic impression of the process. It is unlike proper catalytic processes whereby the overall catalytic reaction is realized first and then the possible elementary processes are searched for. Here the constituent, rather simple, reactions are known. Mechanistic investigations are mainly concerned with the step involving the stoichiometric oxidation of ethylene with $PdCl_2$.

Let us first start by writing the basic equations for the whole process comprising a series of simple stoichiometric reactions that add up to 'catalytic' reactions:

$C_2H_4 + PdCl_2 + H_2O \rightarrow CH_3CHO + Pd + 2\ HCl$	ethylene oxidation
$2\ CuCl_2 + Pd \rightarrow 2\ CuCl + PdCl_2$	palladium reoxidation
$2\ CuCl + 2\ HCl + \frac{1}{2}\ O_2 \rightarrow 2\ CuCl_2 + H_2O$	Cu(II) Regeneration
$C_2H_4 + \frac{1}{2}\ O_2 \xrightarrow{PdCl_2/CuCl_2} CH_3CHO + 218\ kJ/mol$	Overall reaction

Then let us consider the first reaction, which is the stoichiometric oxidation of ethylene. The following experimental results have been established from kinetic investigations:

1. The reaction is inhibited by chloride ion concentration.
2. The reaction is inhibited by hydronium ion concentration; $K_H/K_D = 4.05$, consistent with H^+ release in the rate determining process.
3. Ethanol is not an intermediate since it reacts more slowly than C_2H_4 with $PdCl_2$.
4. The reaction is first order with respect to ethylene and $PdCl_2$.
5. No deuterium is present in CH_3CHO formed if D_2O is used as a solvent. Thus the hydride shift is *intra*molecular.
6. In the oxidation of C_2D_4 in H_2O, $K_H/K_D = 1.07$. The very small isotope effect suggests that no C—H bond-breaking is involved in the rate-determining step.

In the light of these findings, the following mechanism is one of the most plausible mechanisms proposed for the reaction:

(1) Formation of Pd-Olefin complex

$$PdCl_4^- + C_2H_4 \rightleftharpoons PdCl_3(C_2H_4)^- + Cl^-$$

The coordination bonding of ethylene to palladium results in the reduction of the electron-density on the olefin thus making it more susceptible to nucleophilic attack. It must be noted here that Pt(II) reacts readily with olefins. However, it is not as reactive as Pd(II) in the Wacker-type process. This is attributed to the fact that back-donation of electrons from the metal atom to the antibonding orbital of the olefin occurs to a greater extent in the Pt(II) complex than in the Pd(II) one, thus making the Pt-bonded olefin a weaker nucleophilic center than the Pd-bonded olefin. Moreover, the Pd-olefin bond is weaker (due to less back-donation) than the Pt-olefin bond; therefore the activation energy of rearrangement is less for

Pd than for Pt. Finally Pd(II) can expand its coordination sphere more readily than Pt(II) thus lowering the activation energy for nucleophilic attack.

(2) Hydration:

$$PdCl_3(C_2H_4)^- + H_2O \rightleftharpoons PdCl_2(H_2O)(C_2H_4) + Cl^-$$

$$PdCl_2(H_2O)C_2H_4 + H_2O \rightleftharpoons PdCl_2(OH)C_2H_4^- + H_3O^+$$

(3) Nucleophilic addition of a coordinated ligand to the bonded olefin (intramolecular) with simultaneous rearrangement to a metal-alkyl complex.

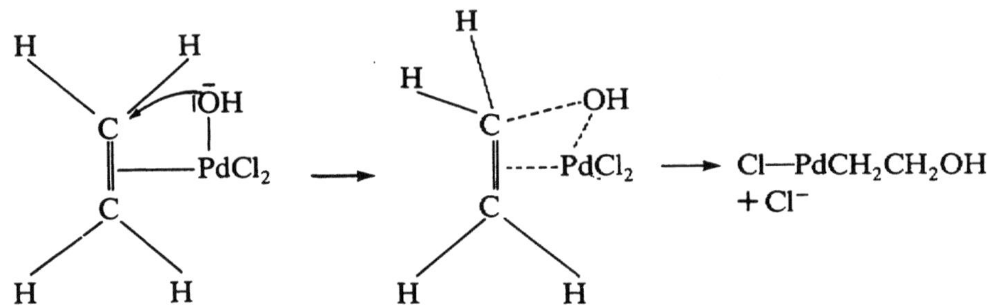

(4) Decomposition of the Pd-alkyl complex to acetaldehyde and Pd(0).

$$ClPd(CH_2CH_2OH) \rightarrow Pd(0) + CH_3CHO + HCl$$

It must be noted that $PdCl(CH_2CH_2OH)$ was prepared independently and shown to decompose to CH_3CHO and Pd(0).

(c) *Acetoxylation* (*Formation of Vinylacetate*)

One of the most interesting catalytic reactions of recent years is acetoxylation, in which a hydrogen atom is replaced by an acetate group in an oxidizing atmosphere with the formation of water. The reaction is applicable to vinyl, allyl, benzyl, and aryl hydrogens. Acetoxylation of ethylene constitutes the major route for the industrial production of vinyl acetate.

Interest in this reaction began with the discovery of Moiseev that vinyl acetate was formed by the interaction of ethylene, acetic acid, a base, and stoichiometric palladium chloride, which was reduced to Pd(0) in the process. Air-oxidation of the reduced palladium, catalyzed by copper ions, can be readily effected, thereby making the reaction catalytic with respect to palladium. Several large-scale plants were built, based on liquid-phase reactions, but they never operated satisfactorily, due to corrosion and other problems. All have been abandoned in favor of vapor-phase processes.

This latter mode of operation eliminates catalyst separation and corrosion problems, and minimizes the formation of acetaldehyde which was always formed as a by-product in the liquid phase process. In fact, the amount of acetaldehyde obtained in the liquid-phase process could be controlled so that the process would provide its own acetic acid, through oxidation of acetaldehyde.

Vapor-phase acetoxylation of ethylene is carried out in fixed-bed tubular reactors over palladium or palladium acetate supported on carriers resistant to attack by acetic acid and containing alkali acetates and other promoters whose function is to increase selectivity, rate and life of the catalyst, and minimize over-oxidation of ethylene to carbon dioxide.

The ratio of reactants is adjusted to operate outside of the explosion limits of about 10% oxygen. This requirement on oxygen concentration limits the per pass conversions to about 25% based on acetic acid (10—15% ethylene and 60—90% oxygen). Selectivities of 91—94% based on ethylene or 98—99% based on acetic acid were obtained. Unreacted ethylene, acetic acid and oxygen are recycled after recovery. The overall reaction may be represented by the following equation. Pure oxygen, rather than air, is used to aid in recovery of ethylene and to eliminate the need for large nitrogen purge.

$$CH_2{=}CH_2 + HOC(=O)CH_3 + \tfrac{1}{2}\,O_2 \xrightarrow[140-180^{\circ}C,\ Pd]{5-10\ bar} CH_2{=}CHOC(=O)CH_3 + H_2O + 176\ kJ/mol$$

The heat of the exothermic reaction is removed by cooling water on the shell side of the tubular reactor and is recoverable at low heat levels. Corrosion problems are not severe and stainless steel reactors can be used. Catalyst life has exceeded 2.5 years. The temperature is set as high as possible, consistent with good selectivity and life, and the pressure is raised as high as possible without approaching the dewpoint. Higher pressures are particularly useful in separation of unchanged ethylene, in increasing productivity, and in lowering the energy requirement for gas recycle.

Catalyst stability depends on operating conditions. It has been suggested that the formation of palladium acetate is the cause of palladium aggregation with its consequent loss in activity. Catalyst activity is therefore maintained by imposing conditions under which palladium acetate does not exist, that is, by lower partial pressures of acetic acid and oxygen, higher temperatures, and higher potassium acetate concentrations.

Although the mechanism of the liquid-phase homogeneous acetoxylation reaction has been intensively investigated, no plausible mechanism has yet been substantiated for the heterogeneous vapor-phase reaction.

An alternative route to vinylacetate has been developed based on methanol and synthesis gas.

$$CH_3OH + 3\ CO + 3\ H_2 \xrightarrow{Cat.} CH_2{=}CHOAc + 2\ H_2O$$

The potential of the industrial commercialization of this process depends on the relative availability of ethylene and methanol.

Vinyl acetate is totally consumed in the manufacture of polyvinyl acetate (used in emulsion paints, adhesives and coatings) and polyvinyl alcohol (also used in adhesives).

5.2.2. PROPYLENE OXIDATION

Propylene is the second most common olefin feedstock in the petrochemical industry (after ethylene). Its oxidation under various catalytic conditions yields a variety of industrially important products. Thus its oxidation is the choice for industrial processes for the manufacture of acrolein, acrylic acid, acrylonitrile, propylene oxide and, to a lesser extent, acetone.

(a) *Acrolein and Acrylic Acid*

Acrolein is the simplest unsaturated aldehyde. It has no major end use by itself. It is used mainly as an intermediate in the manufacture of acrylic acid, which is mainly used in the form of its esters, such as methyl and ethyl acrylates, to make polyacrylates.

The selective oxidation of propylene to acrolein was first realized by Shell in 1948 using cuprous oxide supported on silicon carbide as a catalyst. This discovery catalyzed the investigation of the use of metal oxides in the oxidation of olefins. Mixed metal oxides have been found to be more selective. They give much higher single-pass yield and are more tolerant of oxygen in the gas phase, resulting in a high-conversion process in which almost all propylene is consumed, thus avoiding costly recycle.

Most acrolein plants today are based on mixed oxides of bismuth and molybdenum often called bismuth molybdates. They invariably contain small amounts of one or more of other elements such as P, Ni, Fe, Mn, alkali metals and alkaline rare earths. Acrylic acid plants generally include two reactors. Acrolein is manufactured in the first reactor and is then fed into the second reactor where it is converted, under milder reaction conditions, into acrylic acid.

In a typical process a mixture of propylene, steam and air or oxygen is passed through a stainless steel tubular fluidized-bed reactor at slightly higher than atmospheric pressure, at about 400°C. Under these conditions olefin conversion is about 90% with selectivity of 78% to acrolein; the by-products being acrylic acid, acetaldehyde, acetic acid and carbon dioxide.

Acrolein and by-product acrylic acid are converted to acrylic acid at 250—300°C over a mixed molybdenum and vanadium oxides catalyst containing one or more of P, Sn, Sb, Te or Co. Conversion to acrylic acid is 95% with 85—90% selectivity. It is obvious that the oxidation of acrolein to acrylic acid is easier than the oxidation of propylene to acrolein.

The overall process for making acrolein and acrylic acid may be represented by the following reactions:

$$CH_2{=}CHCH_3 + O_2 \xrightarrow[\text{Bi, Mo oxides}]{400^\circ C} CH_2{=}CHCHO + H_2O$$

$$CH_2{=}CHCHO + \tfrac{1}{2}\, O_2 \xrightarrow[\text{Mo, V oxides}]{250-300^\circ C} CH_2{=}CHCOOH$$

The mechanism of propylene oxidation to acrolein over mixed oxides catalysts has been intensively investigated. Interest in the mechanism of this heterogeneously catalyzed process is comparable to interest in the Wacker process in homogeneous catalytic systems. Although some parts of the process are still not clearly understood, there is strong evidence for the following facts:

(a) The complete (deep) oxidation of hydrocarbons generally takes them to the ultimate products: carbon dioxide and water, via partially oxidized (often industrially desired) intermediates. Acrolein is such an intermediate in the present case. Catalysts of restrained activity are required to intercept the oxidation process and stop it at the level of the formation of the desired intermediates. The mixed metal oxides are such catalysts in the case of acrolein.

(b) The first step in the oxidation of propylene is its adsorption on the oxide surface. The olefin will attach itself by forming a coordination covalent bond to the metal atom through its filled π and empty π^* orbitals.

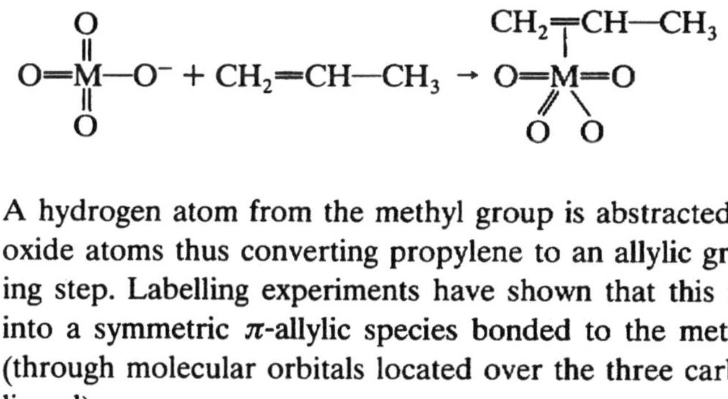

(c) A hydrogen atom from the methyl group is abstracted by one of the adjacent oxide atoms thus converting propylene to an allylic group in a rate-determining step. Labelling experiments have shown that this intermediate rearranges into a symmetric π-allylic species bonded to the metal in a trihapto fashion (through molecular orbitals located over the three carbon atoms of the allylic ligand).

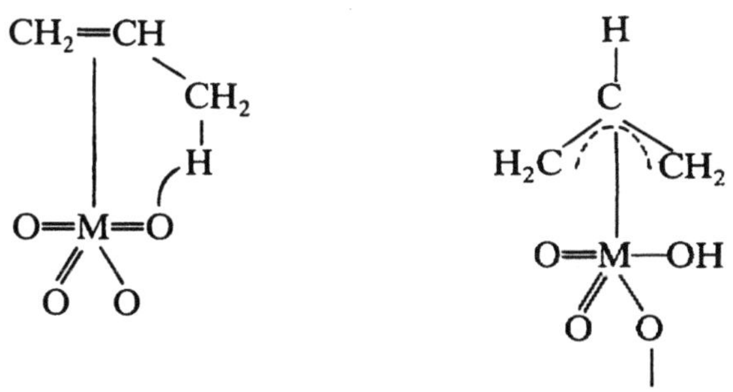

(d) Insertion of a lattice oxygen atom in the metal-allylic bond to form a σ-*O*-allyl species. This is followed by hydrogen abstraction by one of the surface oxygen atoms to yield the aldehyde. This is followed by abstraction of a water molecule from two surface hydroxyl groups. Thus, propylene is oxidized by lattice oxygen of the catalyst. This creates oxide-deficient sites on the catalyst, with the accompanying reduction of the metal.

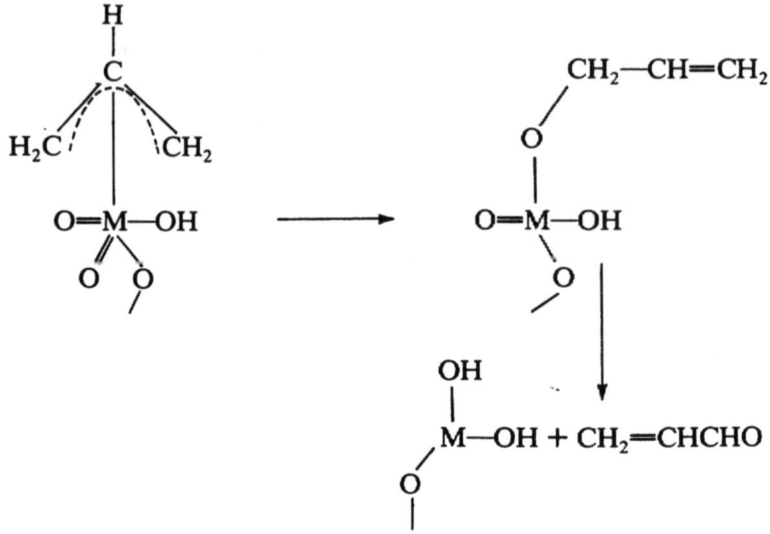

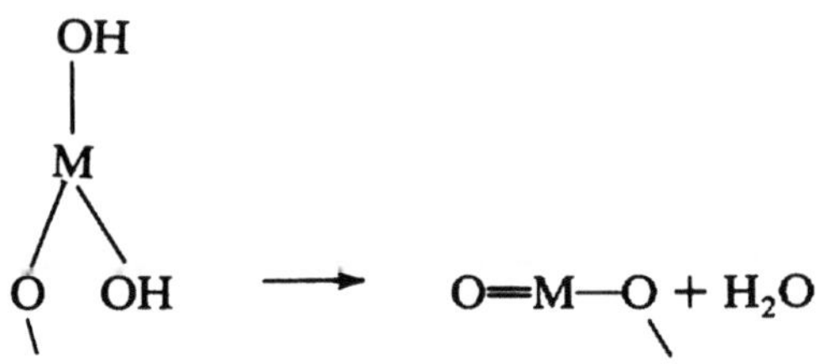

(e) Molecular oxygen is adsorbed on the catalyst surface, gets reduced to the oxide, and migrates to fill the oxide-deficient sites. Thus the catalyst serves as a redox system in a fashion that reminds us of the Pd/Cu system in the Wacker process.

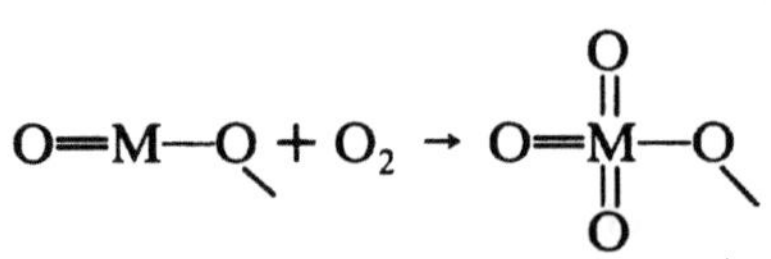

This last reaction is an oversimplification of the facts. Oxygen-deficient sites are not necessarily created and filled over the same metal atom.

All the steps involving oxidation of the allylic species and replenishment of oxygen to the catalyst are fast relative to hydrogen abstraction, leading to the formation of the allylic species. Kinetic investigations therefore do not provide very useful results on the nature of these fast steps and so these steps are still not clearly understood. Evidence substantiating the fact that lattice oxygen is used in oxidizing propylene has been obtained from correlation between detailed structure of these catalysts (structure defects and layer thickness, etc.) and the efficiency of oxidation.

(b) *Ammoxidation (formation of Acrylonitrile)*

Ammoxidation refers to a reaction in which a methyl group is converted to a nitrile on treatment with ammonia and oxygen. An outstandingly successful use of this reaction is in conversion of propylene to acrylonitrile.

Acrylonitrile is an important organic compound used widely as a monomer or comonomer, and as a building block in a variety of organic syntheses. World annual capacity for the production of acrylonitrile exceeds 4.5 million metric tons.

Earlier, acrylonitrile was made by reaction of hydrogen cyanide with acetylene, or ethylene oxide, or acetaldehyde, but these processes have now been replaced by ammoxidation of propylene.

$$CH_3CH{=}CH_2 + NH_3 + 1\tfrac{1}{2}\,O_2 \xrightarrow[2\ \text{bar}]{400-500^\circ C} CH_2{=}CHCN + 3\,H_2O$$

The reaction is usually carried out in a fluidized bed and to a smaller extent in fixed-bed reactors. The catalysts have undergone numerous refinements in the last two decades; the yields of by-products, hydrogen cyanide and acetonitrile, have been decreased, and conversion increased to a point where the reaction is nearly complete on a once-through basis. Bismuth and molybdenum are frequent components of acrylonitrile catalysts, which may also contain a variety of other elements including nickel, cobalt, iron, tin, antimony, phosphorus, and uranium, on supports such as silica.

The catalysts used in ammoxidation are almost identical to those used in partial oxidation of propylene to acrolein. Oxidation of propylene occurs readily at 322°C over Bi—Mo catalysts but, with ammonia present, no propylene conversion occurs until about 402°C, perhaps because of blocking of propylene chemisorption sites by ammonia. The rate-determining step, as in the case of acrolein, is generally considered to be an α-hydrogen abstraction to form an allylic intermediate, a process which equilibrates the ends of the molecule. The observed kinetics, first order in propylene, and zero order in oxygen and ammonia, are consistent with this suggestion. Subsequent steps are less well-understood. A number of mechanistic proposals have been made and compared. The source of nitrogen in acrylonitrile is believed to be lattice nitrogen incorporated in and at the surface of the catalyst as it is purged with NH_3. Nitrogen is continuosuly replenished in the catalyst by NH_3.

(c) *Propylene Oxide*

About 15% of the world propylene production is consumed in the manufacture of propylene oxide which, in turn, is used in the manufacture of propoxylated polyhydric alcohols for urethane foams and for propylene glycol formation.

Attempts to extend the use of ethylene oxide silver catalysts to the direct epoxidation of propylene by vapor phase oxidation have so far resulted in complete oxidation to CO_2 and H_2O. The search is still going on for an effective catalyst for direct epoxidation of propylene.

At present propylene oxide is manufactured by two main industrial processes: the chlorohydrin method and the hydroperoxide method.

In the chlorohydrin method propylene is reacted with an aqueous chlorine solution to produce chlorohydrins which are then dehydrochlorinated to yield propylene oxide.

$$Cl_2 + H_2O \rightarrow HOCl + HCl$$

$$CH_3CH{=}CH_2 + HOCl \rightarrow CH_3{-}CH(OH)CH_2Cl \quad \text{(mainly)}$$

$$2\ CH_3CH(OH)CH_2Cl + Ca(OH)_2 \rightarrow 2\ CH_3\overset{O}{CH{-}CH_2} + CaCl_2 + 2\ H_2O$$

The overall yield of propylene oxide is about 90%. The major drawback of the process is the high consumption of chlorine, an expensive raw material; and the large amounts of wastes (mainly $CaCl_2$) produced. An improvement of the economics of the process could be realized by combining it with a brine electrolysis plant for the production of chlorine. Here $Ca(OH)_2$ could be replaced by NaOH and the by-product NaCl obtained could be recycled to the electrolysis plant.

In the hydroperoxide method an organic oxygen carrier is used to oxidize propylene to propylene oxide. Two oxygen carriers have been successfully utilized: isobutane and ethylbenzene.

Isobutane reacts, in an uncatalyzed reaction, with oxygen to form *t*-butylhydroperoxide which then reacts with propylene in a liquid phase using a molybdenum catalyst to produce propylene oxide and isobutanol.

$$t\text{-}C_4H_{10} + O_2 \xrightarrow[30\ \text{bars}]{120-150^\circ C} t\text{-but-OOH}$$

$$t\text{-but-OOH} + CH_3CH{=}CH_2 \xrightarrow[\text{Mo}]{100-120^\circ C} CH_3{-}\overset{O}{CH{-}CH_2} + t\text{-but-OH}$$

The key to the economic feasibility of this process is the availability of an economic outlet for the by-product *t*-butanol, which is useful as a gasoline component. It is also used in the manufacture of methyl-*t*-butylether (MTBE), an octane number improver that has been used to substitute tetraethyl lead. *t*-Butanol is also an ingredient in methyl methacrylate manufacture and it can be dehydrated to isobutene which may either be hydrogenated to regenerate isobutane or may be used as an alkylating agent in refinery octane improvement.

Ethylbenzene, on the other hand, reacts uncatalytically with oxygen to produce the hydroperoxide which then reacts with propylene to produce propylene oxide and 1-phenylethanol.

$$C_6H_5CH_2CH_3 + O_2 \rightarrow C_6H_5CH(OOH)CH_3$$

$$C_6H_5CH(OOH)CH_3 + CH_2{=}CH{-}CH_3 \rightarrow CH_3\overset{O}{\widehat{CHCH_2}} + C_6H_5CH(OH)CH_3$$

The by-product is dehydrated over TiO_2 catalyst to styrene. So the economic feasibility of this process will depend on the possibility of marketing styrene profitably.

Recently, direct catalytic epoxidation of propylene using hydrogen peroxide over As, B and Mo derivatives, has been claimed to yield propylene oxide at 90% selectivity in a continuous one-step process. Research and development on this and other possible epoxidation processes is continuing very actively in many organizations.

(d) *Acetone*

Acetone is the simplest saturated ketone. It is mainly used in the manufacture of methacrylic acid, methyl methacrylate and other acrylates. It may be manufactured by the homogeneous catalytic oxidation of propylene. In this extension of the Wacker process, the two-stage operation as described above for ethylene is used. The olefin under pressure is absorbed and oxidized by the cupric chloride—palladium chloride catalyst solution at 90—120°C and 9—12 bar. The reacted solution is taken to a second vessel where the pressure is reduced to 1 atm and the ketone stripped from the solution by steam. The catalyst solution is moved to a third vessel where it is reoxidized for reuse by air under pressure.

The rate of the reaction of propylene (and higher olefins) is considerably slower than ethylene. The rate of absorption of pure propylene is about one-third that of ethylene. The yield of acetone is 92—94%. Catalyst solutions are extremely corrosive to ferrous metals, and the use of titanium or other chloride-resistant materials is highly recommended.

Although the manufacture of acetone by the Wacker process is of commercial significance, the current major industrial source of acetone is based on its production as a co-product in the manufacture of phenol from cumene. Another significant source is from the oxidation of 2-propanol.

5.3. Oxidation of Alcohols: Methanol to Formaldehyde

The usual product of the mild oxidation of an alcohol, according to principles of general organic chemistry, is an aldehyde or a ketone which in turn may yield the carboxylic acid upon further oxidation. However, industrial production of aldehydes and ketones is based on alcohols as a feedstock only in a few cases. We have seen that acetaldehyde for example is manufactured by the oxidation of ethylene rather than ethanol.

The only industrially significant process of alcohol oxidation is the oxidation of methanol to formaldehyde.

Formaldehyde is used mainly in the manufacture of phenol-formaldehyde, urea-formaldehyde and other thermoplastic resins.

The dehydrogenation of methanol to formaldehyde is endothermic, the formation of formaldehyde becomes favored at temperatures about 400°C.

When oxygen is added to the system the oxidation reaction takes place:

$$CH_3OH + \frac{1}{2}\,O_2 \rightarrow HCHO + H_2O$$

The reaction becomes exothermic and the removal of water drives the equilibrium in favor of formaldehyde formation. The explosive range of a mixture of methanol and oxygen lies between 36.5% v/v and 6.7% v/v, so the methanol/air feed stream has to be either above the upper limit or below the lower explosive limit.

There are two main industrial methods that are used in the manufacture of formaldehyde from methanol and oxygen. They differ in:

(a) the methanol-to-air ratio, being higher than 36.5% for one (the silver method) and less than 6.7% for the other (iron/molybdenum) method.
(b) the catalyst, being silver metal for one and mixed iron and molybdenum oxides for the other, and
(c) the chemistry.

The choice between the two methods is not straightforward. The silver catalyst, for example, is more expensive and short lived (2—6 months), but may be recovered and regenerated without any loss. The operating cost is higher in the silver method but the capital investment (construction and equipment) is higher in the case of the molybdenum catalyst.

5.3.1. SILVER PROCESS

In this process a mixture of equimolar amounts of methanol and air is passed, slightly above atmospheric pressure, over unsupported silver wire gauze, at about 600°C. The exit gases always contain appreciable amounts of hydrogen and water. The reaction may be considered as a combination of dehydrogenation and partial oxidation:

$$CH_3OH \rightarrow HCHO + H_2 \qquad -84\ \text{kJ/mol}$$

$$CH_3OH + \frac{1}{2}\,O_2 \rightarrow HCHO + H_2O \qquad +158\ \text{kJ/mol}$$

However, intensive investigations of the mechanism of the process suggest that the process does not consist of separate dehydrogenation and oxidation reactions.

The mechanism of the silver process is similar, in some aspects, to the silver-catalyzed epoxidation of ethylene. Here, as in the ethylene process, methanol is adsorbed on the silver surface, on which oxygen has already been chemisorbed. A clean silver surface has no catalytic activity in the absence of oxygen. The following steps constitute the most plausible mechanism:

(i) Chemisorption of O_2:

$$O_2 \overset{Ag}{\rightleftharpoons} 2\,\ddot{O}(ad)$$

(ii) Dissociative adsorption of methanol on O-covered surface:

$$CH_3OH + O(ad) \rightarrow CH_3O(ad) + OH(ad)$$

(iii) Water abstraction from adsorbed hydroxyl group thus providing O(ad) sites.

$$2\,OH(ad) \rightarrow H_2O + O(ad)$$

(iv) Hydrogen abstraction from adsorbed methoxy group to produce free formaldehyde and adsorbed hydrogen atom. This is the rate determining step.

$$CH_3O(ad) \rightarrow HCHO + H(ad)$$

(v) Combination of two adsorbed hydrogen atoms to form a free hydrogen molecule.

$$H(ad) + H(ad) \rightarrow H_2$$

5.3.2. IRON MOLYBDATE PROCESS

In this process the methanol—air feedstream contains 6—9 mol% methanol. The catalyst is generally $Fe(MoO_4)_3$ containing some MoO_3. The reaction temperature is 350—400°C, and it may not exceed 420°C in order to minimize the volatilization of molybdenum oxides. A multitubular reactor is used. The mechanism of the reaction is suggested as being similar to the bismuth molybdate oxidation of propylene to acrolein i.e. the catalyst acts as an oxygen carrier and the oxygen used to oxidize methanol is the lattice oxygen of the catalyst oxides. Molecular oxygen is then reduced to replenish the oxide ions in the catalyst. Labelling experiments have indeed provided results substantiating this fact. However, while it has been established that a mixture of oxides of different metals is more reactive than the pure oxide of any of the components of the mixture in the case of propylene oxidation, the kinetics and mechanism were found to be essentially the same over MoO_3, $Fe(MoO_4)$ or their mixtures in the case of methanol.

The rate-determining step is the C—H bond breaking in the adsorption of the methoxy group.

Effectiveness of metal oxides in converting methanol to formaldehyde has been linked to both their acid-base properties and to their oxidizing ability. Methanol can be attacked at either its basic oxygen or acidic hydrogen. When attacked and activated at the oxygen atom by acidic oxides, such as V_2O_5, MO_3, WO_3, or U_3O_8, formaldehyde is formed, but relatively basic oxides, such as TiO_2, SnO_2, Fe_2O_3, ZnO, NiO, Co_3O_4, attacking the acidic hydrogen produce carbon dioxide. The principle is valid also in composite oxides and formaldehyde will be produced mainly as long as the composite is acidic enough.

However, acidity alone is insufficient, and effective formaldehyde catalysts must have an oxidizing ability as well. Strongly acid catalysts, such as WO_3—P_2O_5 (rich in P), lacking either oxidizing function or basic properties, produce methyl ether rather than formaldehyde.

5.4. Oxidation of Aldehydes

The classical method of making acids is by oxidation of aldehydes. Acids of industrial significance that are manufactured by aldehyde oxidation are: acrylic acid and acetic acid. We have already discussed the manufacture of acrylic acid along with the formation of acrolein by the partial oxidation of propylene.

5.4.1. ACETIC ACID

Acetic acid is the most commonly used organic acid. Its main uses include the manufacture of vinyl acetate, acetic anhydride and esters. It is manufactured by one or more of the following processes:

(a) Non-catalytic oxidation of acetaldehyde which presently accounts for about 70% of acetic acid production.
(b) Oxidative degradation of alkanes and alkenes accounting for about 25% of production.
(c) Carbonylation of methanol, accounting for only 5% of production, but expected to overtake other processes in the future. This will be discussed under carbonylation reactions.
(d) Enzymic oxidation of ethanol. This is used to produce dilute aqueous acetic acid solutions for foodstuff applications.

The current process of making acetic acid from acetaldehyde is based on the oxidation of acetaldehyde by peracetic acid with or without a redox catalyst e.g. cobalt or manganese acetate.

5.5. Oxidation of Aromatic Hydrocarbons

Aromatic compounds are produced in large quantities in refinery processes, such as the reforming of gasoline, or as a by-product of the ethylene/propylene production via steam cracking of naptha.

Table 5-1. Oxidation products of aromatic hydrocarbons.

Product	Capacity in 10^6 t	Starting material	Process Method	Conditions
Phenol	2.0	Benzene	(c)	With H_2SO_4/NaOH
		Benzene	(c)	With HCl/NaOH
		Toluene	(a)	CuO/Air
		Cumene	(a)	Via hydroperoxide with Cu, Co, or Mn salts
Maleic Anhydride	0.5	Benzene	(b)	V_2O_5, 400°C, 1—5 bar
Phthalic Anhydride	2.0	Naphthalene	(b)	V_2O_5/Al_2O_3, 360°C
		o-Xylene	(b)	V_2O_5/TiO_2, 400°C
		o-Xylene	(a)	Co, Mn-acetate 150°C
Terephthalic-Acid	5.6	*p*-Xylene	(a)	Co, Mn salt, 200°C
Anthraquinone	0.03	Anthracene	(a)	CrO_3, 100°C
			(b)	Fe_2O_3/V_2O_5, 350°C

(a) Liquid Phase, (b) Gas Phase, (c) Indirect Oxidation.

Industrial processes for the oxidation of hydrocarbons may belong to one or the other of the following categories:

(a) Liquid Phase Oxidation: This is a homogeneous catalytic process. The catalyst, very often a cobalt salt (e.g. cobalt benzoate), is dissolved in the liquid mixture of the starting materials and oxygen gas is injected into the liquid. It is a relatively mild oxidation method, which works at temperatures up to 200°C and elevated pressures (up to 300 bar). In this process only the side chains are oxidized, the aromatic rings are usually left unchanged.

(b) Gas Phase Oxidation: This is a heterogeneous catalytic process. A solid catalyst, very often V_2O_5, is placed in a tubular reactor and the aromatic hydrocarbon and oxygen are passed over the catalyst at temperatures around 400°C and relatively low pressures of 2—5 bar. The gas-phase oxidation often leads to loss of aromaticity.

(c) Indirect Oxidation: In many cases the direct oxidation processes lead to unwanted side products. Therefore, sometimes an indirect method like oxychlorination or ammoxidation becomes advantageous.

All oxidation processes are highly exothermic and require an efficient cooling system.

Table 5-1 lists the main industrial processes for the oxidation of aromatic hydrocarbons.

In the following sections industrial oxidation processes of aromatic hydrocarbons will be discussed on the basis of the product obtained rather than the process type which may be applicable to the oxidation of more than one aromatic feedstock. Alternative industrial methods for the manufacture of each product will be mentioned even if they did not involve direct oxidation.

5.5.1. PHENOL

Phenol is produced from benzene by indirect oxidation. All attempts to oxidize benzene directly to phenol proved unsuccessful because of the formation of appreciable amounts of by-products like biphenyl and CO_2. The classical process is the benzene sulfonic acid route:

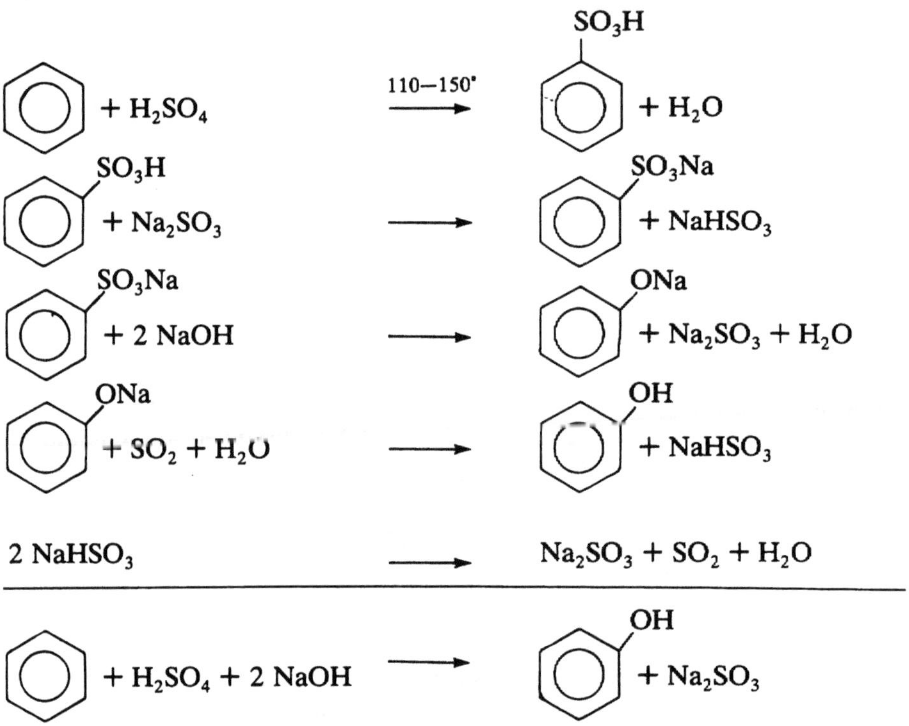

Sulfuric acid is the oxidizing agent, which is consumed during the oxidation process.

In the Raschig-Hooker process chlorobenzene is the intermediate. It is formed from benzene and chlorine, which is produced by *in-situ* oxidation of HCl with a $CuCl_2$ catalyst. The chlorination is then catalyzed by $FeCl_3$. Both catalysts are doped on the same carrier, Al_2O_3:

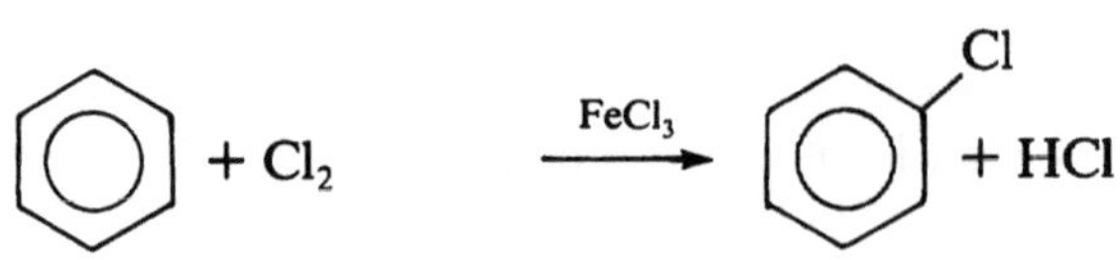

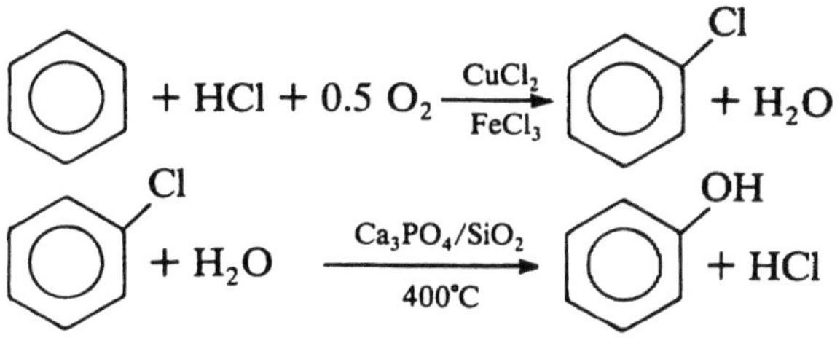

The hydrolysis is also a catalytic process. There are, however, also non-catalytic processes available. Starting from toluene the oxidation is more complex than initially apparent. The first step is the formation of benzoic acid in a typical liquid phase process.

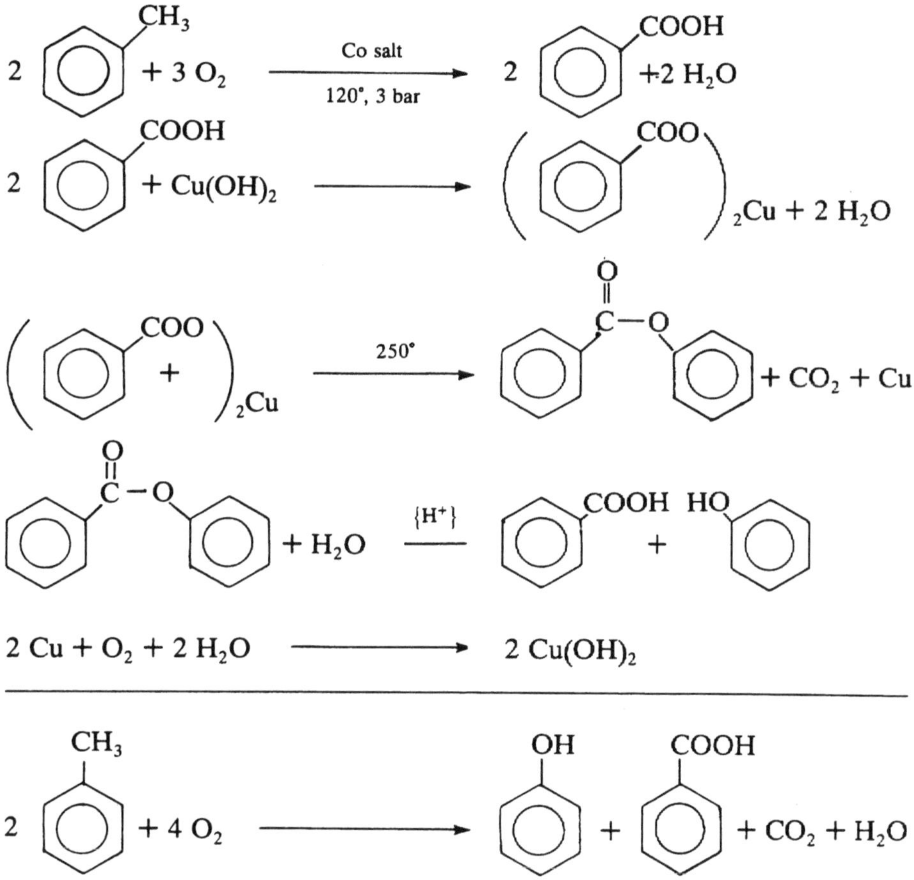

A Cu salt is used to convert 50% of the benzoic acid into phenol. The copper is reoxidized by air and functions as a real catalyst. The product phenol and benzoic acid are both useful salable materials.

Another important process for the production of phenol starts with benzene and propylene. The first step is a Friedel-Crafts alkylation of benzene to produce cumene. This is followed by the actual oxidation step, which is the catalytic formation of the hydroperoxide in a liquid-phase process with cobalt or manganese salts as the catalyst.

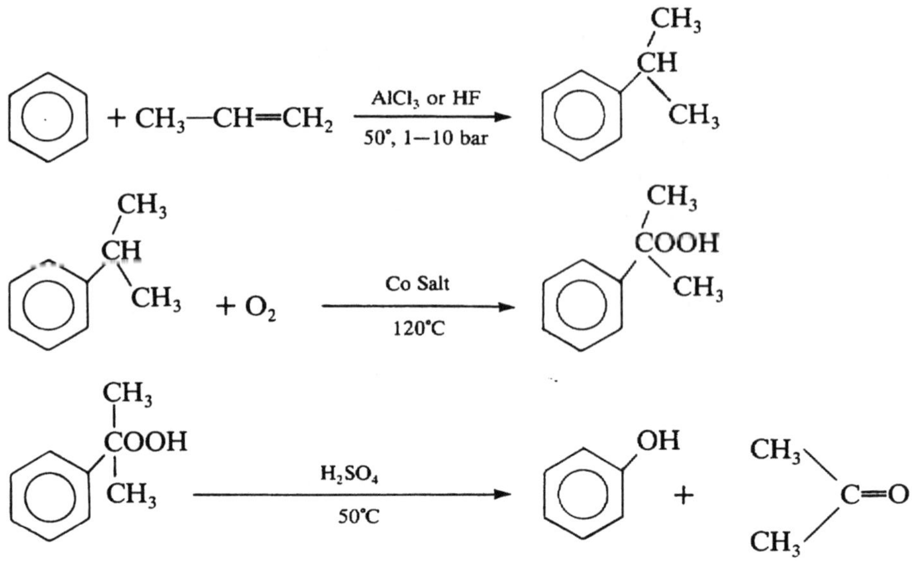

The hydroperoxide forms phenol and acetone in an acid-catalyzed reaction.

Phenols are used for the production of phenol resins, nylons via ε-caprolactam or adipic acid, and epoxy resins and polycarbonates via bisphenol A. Phenol is also an important intermediate for the manufacture of pharmaceuticals, e.g. acetyl salicylic acid (= Aspirin).

5.5.2. MALEIC ANHYDRIDE

Maleic anhydride is produced by the oxidation of butane or benzene. The benzene oxidation is the oldest process and is still viable:

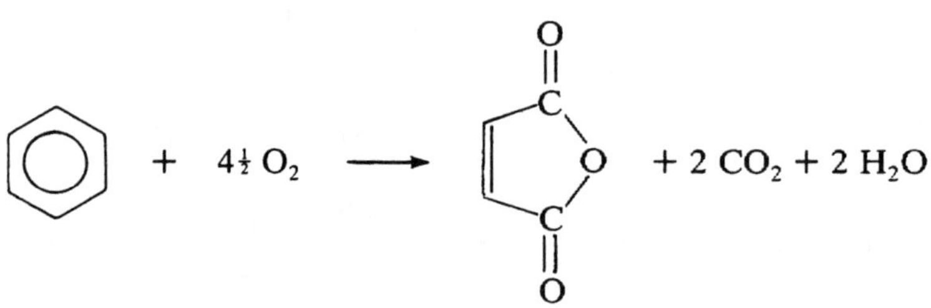

The reaction is heterogeneously catalyzed by V_2O_3/MoO_3 at 380°C and 1—5 bar. It is exothermic with a heat of reaction of 2260 kJ/mol.

Benzene conversion reaches 90% but selectivity to maleic anhydride is only 50—60%, the other 40—50% is completely oxidized to CO_2. The reaction is highly exothermic. Therefore several heat exchangers are required to cool the product.

Maleic anhydride is used for the production of polyesters and as an intermediate in the synthesis of pesticides.

5.5.3. PHTHALIC ANHYDRIDE

About 80% of the world phthalic anhydride is produced by the oxidation of *o*-xylene and 20% by the oxidation of naphthalene. Xylene can either be oxidized in a liquid or a gas-phase process, with the gas-phase process being more abundant.

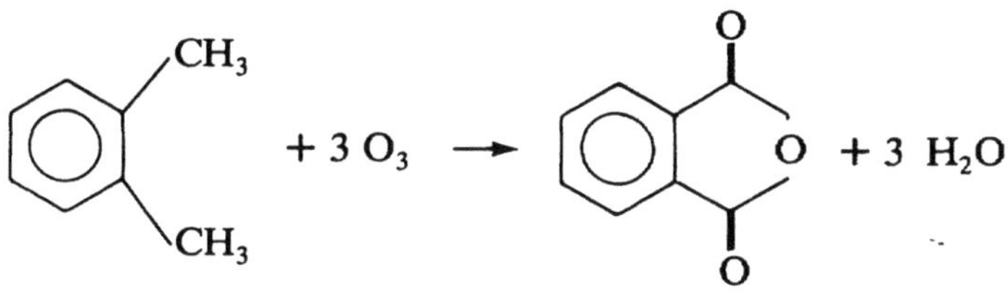

Gas-Phase: V_2O_5 pure or on TiO_2 in a tubular fixed-bed reactor or a fluidized-bed reactor.

Liquid-Phase: Cobalt or manganese acetate in acetic acid as the solvent at 150°C.

For the naphthalene oxidation, the reaction is carried out in fixed-bed or fluidized-bed reactors at 360—380°C with V_2O_5 as a catalyst. The selectivity for phthalic anhydride is around 75%. The major by-products are maleic anhydride and 1,4-naphthoquinone.

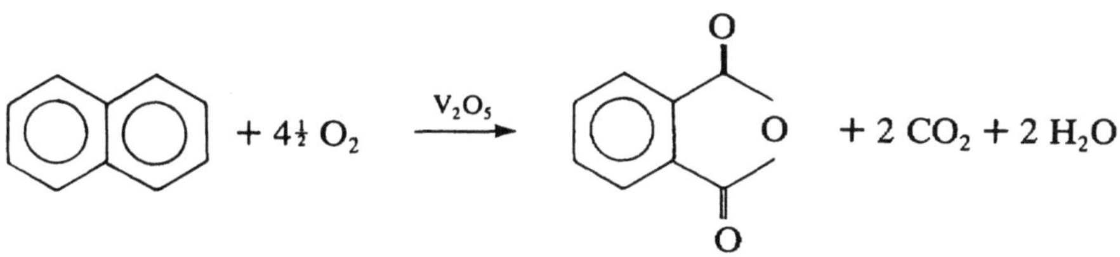

Phthalic anhydride is mainly reacted with C_4—C_{10} alcohols to the corresponding esters. The esters are employed as plasticizers for PVC. The most important ester is DOP (dioctylphthalate), which is obtained via the esterification of phthalic anhydride with 2-ethylhexanol, a major product of the OXO-process.

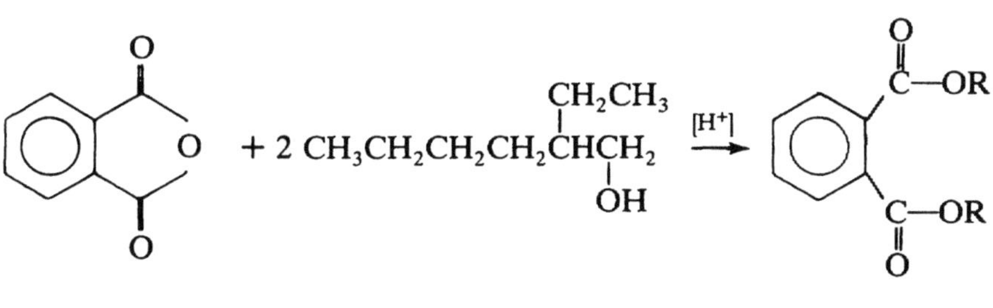

5.5.4. TEREPHTHALIC ACID

Terephthalic acid is produced by liquid-phase oxidation of *p*-xylene. However, if no special precautions are taken the oxidation stops at the stage of the *p*-toluic acid:

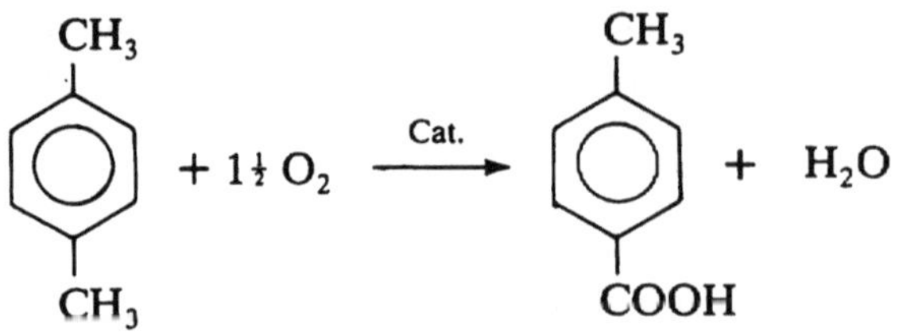

To overcome this problem, several processes have been developed.

(1) Esterification of toluic acid with methanol and oxidation of the remaining methyl group in a separate step.
(2) A more active catalyst is obtained when a promoter, such as bromine, is added to cobalt or manganese salt, which facilitates the oxidation of both methyl groups simultaneously.
(3) In a co-oxidation process an auxillary substance, like acetaldehyde, paraldehyde or methyl ethyl ketone, is added. They form hydroperoxides under the reaction conditions, which initiate the oxidation of the second methyl group.

Commonly the oxidation processes are carried out in methanol as the solvent. Therefore dimethylterephthalate is the primary product, which must be hydrolyzed to give the pure acid.

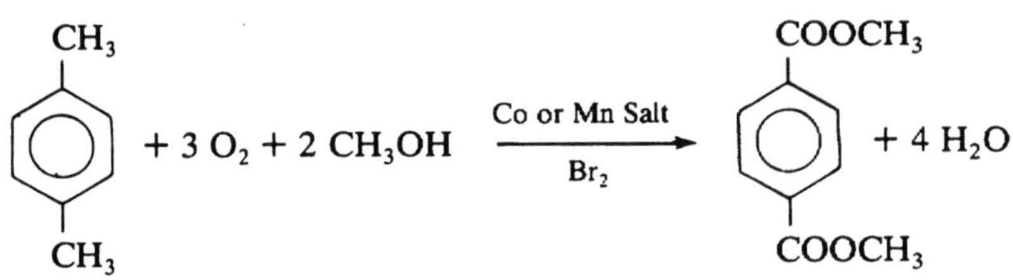

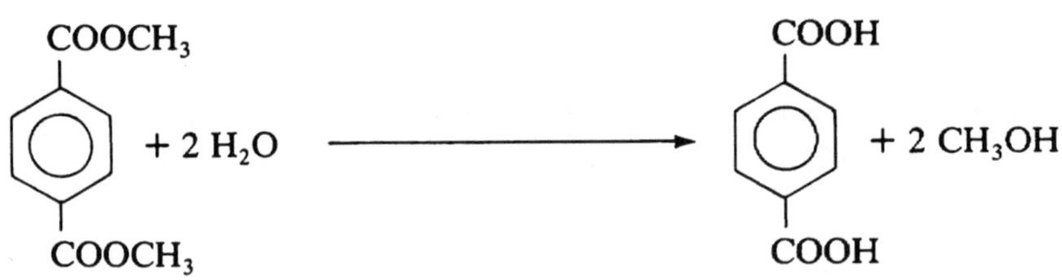

Terephthalic acid and dimethylterephthalate are starting materials for the manufacture of polyesters whose chief application is in the fiber sector. Reaction with diols, like ethylene glycol, leads to polyethylene terephthalate (PET), the most important polyester.

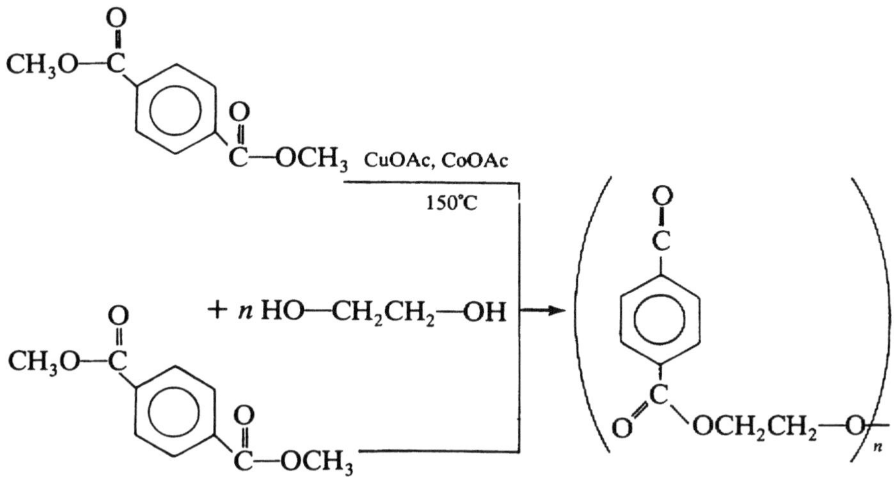

5.5.5. ANTHRAQUINONE

Although the world production of anthraquinone is less than 50,000 t/year, it is important, since it is used as an intermediate for the manufacture of high value dyes.

The oxidation is conducted in the liquid phase with CrO_3 at 100°C or in the gas phase with iron vanadate as the catalyst at 380°C.

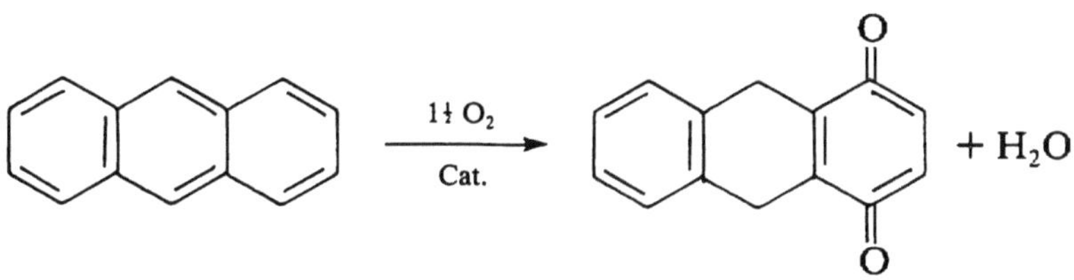

Anthraquinone itself is used as an organic catalyst for the production of hydrogen peroxide, in a quite unique process (originally by IG Farben, now by DuPont):

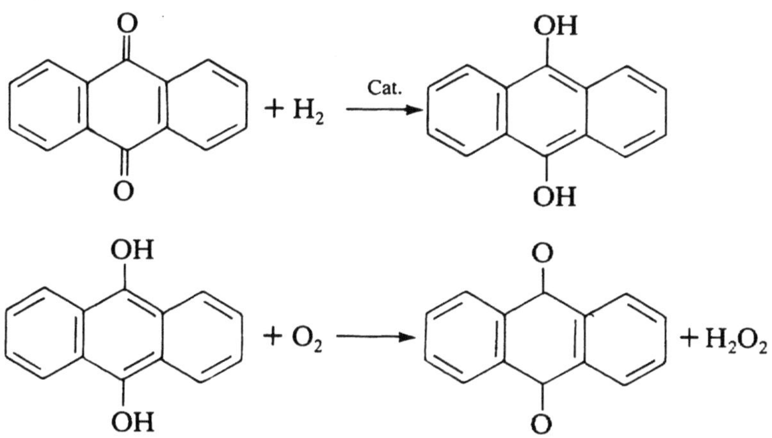

5.5.6. AMMOXIDATION

Instead of using oxygen as an oxidizing agent a mixture of O_2 and NH_3 may be used to oxidize xylenes. The products are nitriles. This ammoxidation is carried out in the gas phase with bismuth molybdate or V_2O_5/Al_2O_3 as the catalysts at a temperature around 400°C in a process similar to the ammoxidation of propylene.

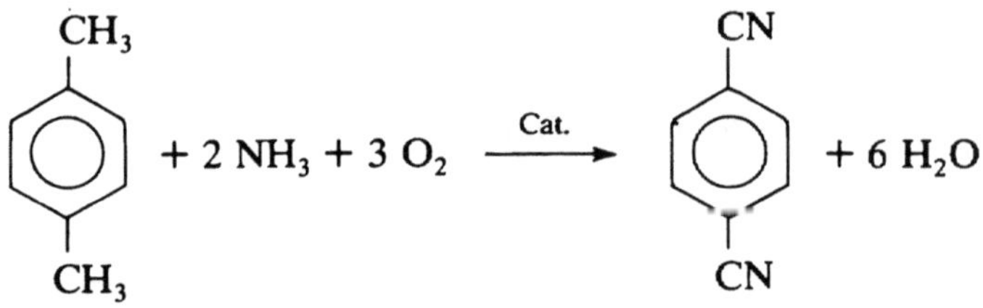

The nitriles are used to manufacture the corresponding acids or diamines, or as intermediates.

References

Gates, B. C., Katzer, J. R., and Schuit, G. C. A.: *Chemistry of Catalytic Processes*, McGraw-Hill Book Co., 1979.

Grasseli, R. K. and Burrington, J. D.: 'Selective Oxidation and Ammoxidation of Propylene by Heterogeneous Catalysis' in *Advances in Catalysis*. Edited by Eley, D. D., Pines, H., and Weisz, P. B., Academic Press, 1981, pp. 133–165.

Pearce, R. and Patterson, W. R.: *Catalysis and Chemical Processes*, John Wiley and Sons, 1981.

Satterfield, C. N.: *Heterogeneous Catalysis in Practice*, McGraw-Hill Book Co., 1980.

Voge, H. H. and Adams, C. R.: 'Catalytic Oxidation of Olefins', *Adv. in Cat.* **17**, 151–221 (1967).

Sheldon, R. A. and Kochi, J. K.: 'Metal-Catalyzed Oxidation of Organic Compounds in the Liquid Phase: A Mechanistic Approach', *Adv. in Cat.* **25**, 272–413, 1976.

CHAPTER 6

Production and Uses of Synthesis Gas

6.1. Introduction

The term 'synthesis gas' is used for gas mixtures which are feedstocks for the synthesis of chemical intermediates, such as methanol, OXO aldehydes, etc. It usually denotes a mixture of carbon monoxide and hydrogen containing small amounts of CO_2 and methane.

Synthesis gas is the most important chemical feedstock. It may be used as a mixture of CO and H_2 in the proper ratio, as in methanol synthesis and hydroformylation processes, or as a source of pure CO, as in acetic acid synthesis from methanol and in carbonylation reactions, or as a source of pure hydrogen in ammonia synthesis and hydrogenation reactions.

Synthesis gas is produced by conversion of methane, naphtha crude oil fractions or coal with water. Depending on the feedstock different ratios of CO/H_2 are obtained:

From Coal:

$$C + H_2O \rightleftharpoons CO + H_2 \qquad \Delta H = +130 \text{ kJ/mol} \tag{1}$$

From Oil Fractions:

$$\text{—}CH_2\text{—} + H_2O \rightleftharpoons CO + 2\,H_2 \qquad \Delta H = +151 \text{ kJ/mol} \tag{2}$$

From Natural Gas:

$$CH_4 + H_2O \rightleftharpoons CO + 3\,H_2 \qquad \Delta H = +205 \text{ kJ/mol} \tag{3}$$

These processes are all endothermic, thus requiring energy which can be supplied by an external heating source or by partial combustion of the starting material.

$$\text{—}CH_2\text{—} + \tfrac{1}{2}\,O_2 \rightarrow CO + H_2 \qquad \Delta H = -92 \text{ kJ/mol} \tag{4}$$

$$\text{—}CH_2\text{—} + 1\tfrac{1}{2}\,O_2 \rightarrow CO_2 + H_2O \tag{5}$$

The reaction of hydrocarbons with water (steam) is a catalytic process and is called *Steam-reforming*. It is mainly used for light feeds, such as natural gas or naphtha. The autothermal cracking process, also known as partial oxidation, is a non-catalytic process. It is preferably, but not exclusively, used for heavy feeds, such as heavy fuel oil or crude oil.

The production of synthesis gas from coal is commonly known as *Coal-gasification*. It is based on partial oxidation of coal and is essentially non-catalytic.

The fraction of hydrogen in the synthesis gas can be increased by the *Water gas shift* reaction. This is the catalytic conversion of carbon monoxide and water to carbon dioxide and hydrogen:

$$CO + H_2O \rightarrow CO_2 + H_2 \qquad \Delta H = -42 \text{ kJ/mol} \qquad (6)$$

6.2. Manufacture of Synthesis Gas

6.2.1. STEAM-REFORMING (ICI Process, based on Schiller process)

The main reaction takes place in catalyst-filled tubes at 700—830°C and 15—40 bar. The reforming tubes are heated by combusting natural gas or other ash-free fuels. The catalyst is Ni supported on α-alumina, containing potassium oxide to accelerate carbon removal reactions.

At the high pressure in the primary reactor the synthesis gas contains some methane, since equilibrium (3) is partly shifted to the left. To get a complete conversion to CO and H_2 a higher temperature is required. However, due to construction material constraints, temperature higher than 830°C cannot be reached in externally heated reforming tubes. If a methane-free synthesis gas is required a second reactor is added in which a small part of the gas is combusted to reach a temperature of over 1200°C. The reactor is filled with a high-temperature-resistant catalyst, which decreases the methane content to less than 0.2—0.3 vol %.

The steam reforming catalysts are readily poisoned by sulfur compounds. If the feed contains sulfur, a hydrodesulfurization unit precedes the synthesis gas production. All sulfur is converted to H_2S by catalytic hydrogenation of the feed on a CaO—MoO_3 catalyst at 400°C. The H_2S is then absorbed on ZnO.

6.2.2. PARTIAL OXIDATION

Partial oxidation differs from steam reforming in two main aspects: it is non-catalystic, and the heat necessary for the process is generated by combustion of a part of the feedstock *in situ* (as the case of the second reactor in steam reforming). The process is applicable to various feedstocks ranging from natural gas to heavy oil. The process does not suffer from catalyst poisons since it is non-catalytic. A supply of oxygen should be available. The reaction mixture, consisting of hydrocarbons, water and stoichiometric amounts of oxygen is led into the reactor at 1200—1500°C and 30—80 bar.

6.2.3. WATER GAS SHIFT REACTION

Two processes to convert CO and water to CO_2 and H_2 are known. The high temperature shift conversion uses a Fe_3O_4—Cr_2O_3 catalyst and operates at 350—400°C. Large excess of steam should be present in the system to prevent reduction of Fe_3O_4 to metallic Fe, which could catalyze the formation of carbon from carbon monoxide. Under these conditions the reaction is fast, but at equilibrium it still contains about 3 mol% CO. If a complete CO conversion is required, a more effective low temperature catalyst is employed in a second unit. CuO—ZnO catalyses the shift conversion at 190—260°C, temperatures at which the equilibrium contains only 0.1% CO. Copper metal is the active catalytic species. Zinc oxide serves to protect Cu from poisoning by adsorbing traces of sulfur compounds. Zinc oxide also acts as a support for copper.

6.2.4. SYNTHESIS GAS PURIFICATION

If the synthesis gas is to be used in catalystic processes, it must not contain acid gases like H_2S, COS and (sometimes) CO_2, since they poison the catalysts. Therefore the synthesis gas is washed with pressurized methanol (Rectisol by Lurgi), dimethylglycol ethers (Selexol by Allied), sulfolan + diisopropylamine (Sulfinol by Shell), or monomethanolamine (MEA) aqueous solution. Traces of CO_2 and CO may be further removed by catalytic reduction to methane with hydrogen over supported Ni at 250—350°C.

These reactions are the reverse of steam-reforming. Levels of carbon oxides can thus be reduced to a level of a few ppm; an important requirement in ammonia synthesis and organic hydrogenation where hydrogen should be ultrapure.

$$CO_2 + 4\,H_2 \rightleftharpoons CH_4 + 2\,H_2O$$

$$CO + 3\,H_2 \rightleftharpoons CH_4 + H_2O$$

Table 6-1. Economics of synthesis gas production.

	Steam Natural Gas	Reforming naphtha	Partial Oxidation of residual oil	Gasification of coal
Pressure (Psig)	318	318	14.50	318
H_2/CO (mol)	5/1	5/1	2.7/1	5/1
Capital Cost (Million $)	24.8	28.8	71.7	115.9
Labor Cost	2.1	2.3	6.9	9.2
Feed Cost	54.4	60.5	77.9	35.7
Catalyst etc.	1.6	2.2	4.5	5.8
Utilities	36.2	37.4	10.7	25.1
Indirect Cost	11.2	12.9	31.8	49.6
Total Cost	105.6	115.6	131.8	126.4
Credit Steam	−9.8	−7.7		
Net Production Cost	95.8	107.9	131.8	126.4
Base:	120 millions scfd synthesis gas, 329 days/year			

Source: Process Economics Program Report No 110, 'Synthesis Gas', SRI, Menlo Park, Calif. 94025.

6.2.5. ECONOMICS OF SYNTHESIS GAS PRODUCTION

Comparison of the economics of the various processes for making synthesis gas (Table 6-1) shows that methane (natural gas) has advantages over the other feedstocks. However, in some areas in which natural gas is rare and expensive, naphtha is used as the feedstock. Residual oil is essentially a waste product, which must be disposed of. Therefore, the cost factor is less important, since the production of synthesis gas is an environmentally favored method to convert the oil into a useful intermediate (synthesis gas).

6.2.6. THE CATALYSTS

The mechanism of the steam reforming process is not completely understood. It seems, however, that Ni catalysts mainly facilitate the cracking and dehydrogenation of the hydrocarbon feed:

$$CH_4 \xrightarrow[850°C]{Ni} C + 2\,H_2$$

The alkali oxide (part of the catalyst), due to its basic sites, absorbs water and enhances the carbon steam reaction:

$$C + H_2O \xrightarrow{K_2O} CO + H_2$$

The active part in the high pressure shift catalyst is believed to be Fe_3O_4, that is, a combination of Fe(II) and Fe(III) oxides. The shift catalyst is prepared by mixing aqueous solutions of $Fe(NO_3)_3$ and $Cr(NO_3)_3$. Addition of ammonium oxalate leads to a precipitate of iron and chromium oxalate. The precipitate is carefully washed with water to remove ammonium and nitrate ions. The precipitate is dried at 150°C and then calcined at higher temperature (Ca. 600°C). The product is a mixture of microcrystallites of Cr_2O_3 and Fe_2O_3 with a size of some 1000 nm. Before use, the catalyst is exposed to a stream of hydrogen to reduce the Fe_2O_3 to Fe_3O_4, the actual active phase.

The preparation of the other catalysts is similar. However, companies usually keep the exact preparation procedure a secret, since very often the mode of preparation affects the activity of a catalyst more than its composition. This is due to the fact that the internal surface area, pore size distribution, size of crystallites, etc., in a catalyst are very important for its performance.

The performance of conventional alkali-promoted catalysts tends to be unsatisfactory as ever-higher pressures and temperatures are being applied in modern steam reforming. This is particularly true of naphtha reforming where carbon deposits may form, if a sufficiently high steam flow is not provided. Under these conditions, however, the alkali tends to migrate downstream, leaving inactive catalyst sites and depositing alkali in cooler sections of downstream equipment. The resultant drop in catalyst activity causes local hot spots and eventual tube failure through overheating. Carbon can be deposited in the interstices of the catalyst particles, setting up stresses that may physically destroy the catalyst. To overcome these difficulties a number of catalyst formulations have been designed that eliminate the need for alkali additives or improve the physical durability of catalyst particles through the use of various catalyst supports.

ICI has found that the use of ruthenium and calcium oxide on an alumina support results in a catalyst that is more active than the conventional alkali-promoted one. The Gas Council, on the other hand, suggests the use of uranium oxides as the promoter, with nickel oxide content (12% w/w) lower than in conventional types which makes it a more active and stronger catalyst. BASF, too, has found that lower nickel content (2—6%) on α-alumina, with some activated alumina, produces a more durable catalyst with higher activity.

Both Produits Chimiques and Mitsubishi Chemical have observed that the addition of tungsten to low nickel content alumina tends to produce a catalyst with a lower tendency to cause carbon deposition.

One manufacturer had good success in changing from natural gas feed to naphtha by replacing the catalyst in the reformer.

6.2.7. DESCRIPTION OF THE PROCESS OF STEAM REFORMING OF NATURAL GAS

A simplified flow sheet for steam reforming process is shown in Figure 6-1. The main components are: desulfurization unit, reformer (the furnace), heat recovery unit and synthesis gas purification unit. Natural gas is first desulfurized in one of two activated carbon beds to the less than 0.2 ppm vol sulfur required for the reformer catalyst. Each carbon bed is regenerated with steam after a three day operating cycle. The desulfurized natural gas is then mixed with saturated steam (steam : carbon ratio 4 : 1) and preheated in the first coil of the convection section of the reforming furnace to approximately 550°C. The radiant section of the vertical tube reformer furnace is fired by fuel oil using 10% excess air which has been preheated to 180°C. The reformer combustion gases, as well as being used to preheat the reformer feed, also serve to preheat the air in the regenerative air preheater and to raise high-pressure saturated steam in the waste heat boiler. Flue gases exit the reformer at 165°C and are sent to the stack via a forced draft blower.

Raw product synthesis gas leaves the reformer at 22 bar and is first used to generate steam in the waste heat boiler and to preheat boiler feedwater before supplying the heat for the purification section stripper-reboiler. The synthesis gas then undergoes further cooling by heat exchange with boiler feedwater and cooling water. The condensate from the synthesis gas is returned to the deaerator for use as feedwater to the steam generating facilities.

The synthesis gas, now cooled to 50°C, is treated in a regenerative mono-ethanolamine (MEA) purification system for the removal of carbon dioxide down to a level of 0.1% V. MEA in a 35% w/w aqueous solution is used as the absorption medium. The carbon dioxide-loaded solvent is sent from the bottom of

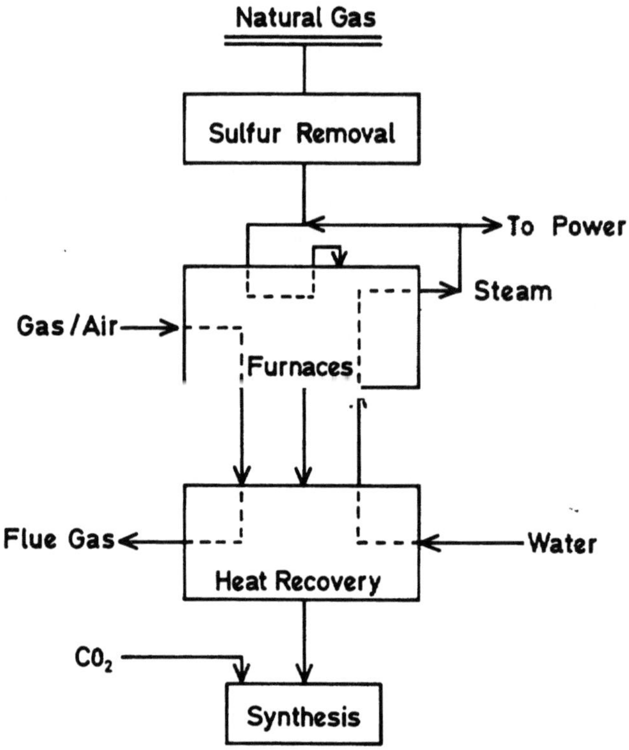

Fig. 6-1. Natural gas reforming for production of synthesis gas.

the absorber to a stripping column where the solution is regenerated at nearly atmospheric pressure by means of heat supplied to the reboiler. The regenerated MEA solution, after heat exchange with the carbon dioxide-rich solution, going to the stripper, is pumped to the absorber pressure, cooled to approximately 38°C, and is returned to the top of the absorber. Acid gas, primarily carbon dioxide, from the top of the MEA stripper is vented to the atmosphere, while the treated synthesis gas leaves the absorber at 21 bar.

A typical composition of synthesis gas obtained from two different feedstocks is given in Table 6-2.

Table 6-2. Composition of raw synthesis gas from two different sources.

Component	Steam Reforming of Methanol % v/v	Heavy Oil Oxidation % v/v
CO	15.5	47.0
CO_2	8.1	4.3
H_2	75.7	46.0
$N_2 + A$	0.2	1.4
CH_4	0.5	0.3
H_2S	0	1.0

6.3. Synthesis Gas as a Chemical Feedstock

Synthesis gas is one of the most important building blocks in the petrochemical industries. The outstanding research efforts that have been put in converting synthesis gas into useful chemical intermediates have paid back tremendous successes. One cannot compare the importance of synthesis gas as a petrochemical building block with methanol or ethylene because synthesis gas itself is the main precursor for methanol and probably will soon be the main source for ethylene too. There is no end to the list of chemical products and intermediates, of industrial importance, that can or will be feasibly made from synthesis gas. The major products that account for its consumption today include:

(1) Ammonia, which accounts for about 80% of natural gas consumption as a petrochemical feedstock. Synthesis gas is used in the manufacture of ammonia as a source of hydrogen.
(2) Methanol is the second major user of synthesis gas, where the latter provides both hydrogen and carbon monoxide.
(3) Oxo chemicals which are mainly produced by carbonylation reaction of olefins to yield aldehydes that are further hydrogenated to alcohols. Synthesis gas here acts as source of carbon monoxide. It should be mentioned here that a special carbonylation reaction of paramount importance in the petrochemical industry is the carbonylation of methanol to yield acetic acid. This is becoming the most important process for the manufacture of acetic acid.
(4) Fischer-Tropsch Chemicals and Fuels. This is a significant user of synthesis gas. However, due to its historical development and its association with coal and heavy oil gasification, the source of synthesis gas in these reactions is mainly coal or heavy oil residues.
(5) Hydrogenation products of petrochemical and refining processes.

Synthesis gas is used as a source of hydrogen in these reactions.

In the following sections we are going to discuss three major uses of synthesis gas: methanol synthesis, Fischer-Tropsch and related syntheses and synthesis of ammonia and the related urea production.

6.3.1. METHANOL SYNTHESIS

The importance of methanol is due to its potential use both as fuel and as a feedstock for the petrochemical industries. Methanol is cheaper to ship than liquified natural gas since it can be transported in conventional crude oil tankers. Its storage is also cheaper and its handling is more convenient and safer. Several new petrochemical processes have been introduced recently based on methanol as a starting material. It has a great potential for more processes that are being either developed or researched.

World production of methanol in 1981 was about 12 million tons and is expected to exceed 20 millions t/a in 1990. The present and potential industrial uses of methanol include its carbonylation to acetic acid, its conversion to gasoline by the Mobil methanol-to-gasoline process, its oxidation to formaldehyde, its use as one of the ingredients of the gasoline additive methyl-*t*-butyl ether (MTBE), its potential use by itself as an octane enhancer, its catalystic cracking to olefins and its homologation to acetaldehyde and ethanol.

Methanol can be synthesized using both homogeneous and heterogeneous catalysts. Some processes and reaction conditions are listed in Table 6-3.

Table 6-3. CH_3OH manufacture.

	Catalyst	T°C	Pressure (bar)
Nissui-Topsoe	$CuO—ZnO—Cr_2O_3$	230—260	100—150
Vulcan	$ZnO—Cr_2O_3$	270—330	150—250
Pritchard	CuO	200—330	100—250
Catalyst and Chemical Inc.	$CuO—ZnO—Al_2O_3$	240—250	100—250
BASF	$CuO—ZnO—Al_2O_3$	200—350	50—250
ICI	$CuO—ZnO—Al_2O_3$	240—260	50—100
Lurgi	CuO—ZnO	230—250	40—50

Commerical methanol catalysts generally consist of Zn and Cu oxides, either alone or together plus a promoter which is usually chromium oxide.

All the processes for synthetic methanol involve the basic steps of:

(1) steam reforming of natural gas plus addition of CO_2 to adjust the C—H ratio,
(2) compression;
(3) synthesis in a catalytic converter and
(4) distillation for purification.

Synthesis gas which is usually a mixture of CO, H_2 and to less extent CO_2 is converted to methanol through the following reactions:

$$CO + 2\,H_2 \rightarrow 2\,CH_3OH$$

$$CO_2 + 3\,H_2 \rightarrow CH_3OH + H_2O$$

Since the optinum ratio for C-to-H is 1 : 2, and natural gas is mostly CH_4 with 1 : 4 ratio, CO_2 is normally added to take care of the surplus hydrogen in the synthesis gas mixture.

The single most important development in methanol synthesis is ICI's new methanol process using a new active Cu catalyst. The process operates at 50 bar and 240°C. It is possible to operate this process on H_2-rich syngas without having to add CO_2. Methanol plants can be built in large size and today's standard plants amount to 2000 MTD.

Three mechanisms have been proposed for the heterogeneously catalyzed synthesis of methanol.

(1) Successive hydrogenation of chemisorbed CO:

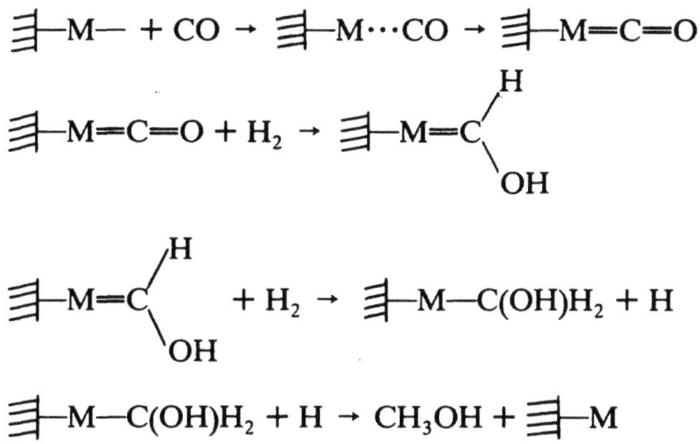

(2) Insertion of CO into M—H bond to form a formyl intermediate, followed by successive hydrogenation:

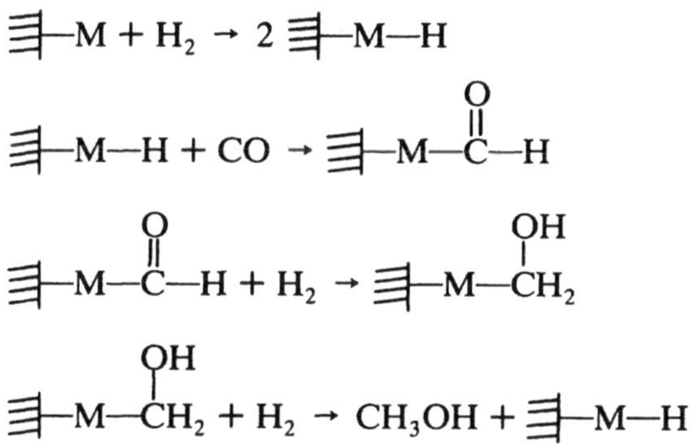

(3) Insertion of CO into a surface O—H to form a surface formate which undergoes hydrogenation followed by consecutive dehydration and hydration to yield methanol:

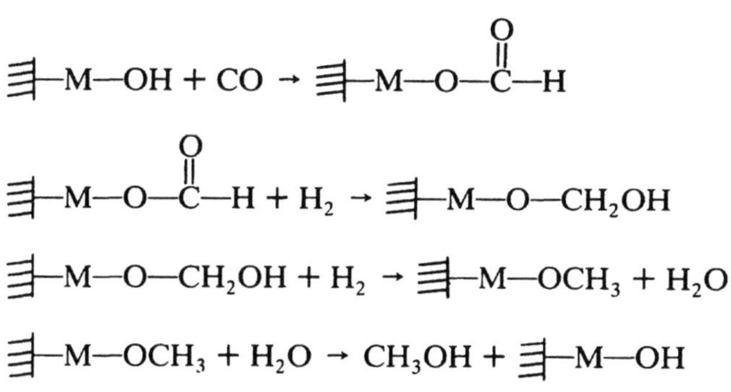

The water molecule abstracted from the adsorbed species upon hydrogenation apparently is not completely desorbed; it remains close enough to the metal site involved in catalysis to release methanol from the adsorbed methoxy group and regenerate the active ≡M—OH site.

Although each mechanism is based on some experimental evidence, there is no conclusive evidence that one of them is the mechanism for the reaction. More than one scheme may also be operating simultaneously.

6.3.2. FISCHER-TROPSCH AND RELATED SYNTHESES

Fischer and Tropsch (1925) reported the catalytic hydrogenation of carbon monoxide over iron-based catalysts at atmospheric pressure to produce a Flory-Schulz distribution of alkanes and olefins. The process was originally intended for the production of synthetic liquid fuels. However, it can very well be applied to the manufacture of synthetic chemicals either by distillation of the synthetic fuels followed by cracking the proper fractions to olefins; or by direct conversion of synthesis gas to lower olefins by applying the proper reaction conditions (catalyst, temp, press, CO/H_2 ratio, reactor design etc.). The process has been developed over several decades and many improvements have been introduced in the various components of the system.

By manipulation of the reaction conditions, the process may be designed to produce heavier saturated hydrocarbons or lower olefins or oxygenated hydrocarbons as we shall see in the following discussion.

Catalysts

Metals that have significant activity for Fischer-Tropsch synthesis include iron, cobalt, nickel and ruthenium. Iron has proved so far to be the best. It is superior to cobalt with respect to conversion rate, selectivity and flexibility. Nickel has the disadvantage of producing appreciable amounts of methane. Ruthenium enhances the formation of high molecular weight alkanes and catalyzes polymerization to polymethylene. Other Group VIII metals are of low activity. Copper does not catalyze Fischer-Tropsch synthesis.

The catalyst is usually prepared by fusion or precipitation over a silica, alumina or kieselguhr support. Small amounts of promoters such as alkali metal or copper salts are included in the catalytic mix. Copper is believed to facilitate the reduction of the catalyst, alkali metal salts, particularly K_2O, enhance activity and olefin selectivity. The support increases the surface area of the catalyst. Some supports e.g. TiO_2 may interact strongly with the catalyst metal thus extremely increasing its dispersion.

Sulfur compounds generally poison the catalyst and they must be removed from the synthesis gas feedstream. However, partial sulfur poisoning may have favorable effects. Thus, it has been found that deliberate slight sulfur poisoning of the iron/manganese catalyst enhances selectivity to short chain olefins.

Reactors

Three main types of reactors are currently in use in the Fischer-Tropsch processes: fixed-bed, fluidized-bed and slurry bed reactors. Fixed-bed reactors are usually tubular, each tube having 50 mm ID and 12 m length. A single reactor may contain as many as 2000 such tubes. Fluidized-bed reactors provide for better heat transfer and continuous regeneration of the catalyst. The catalyst used in fluidized-bed reactors must have high physical stability. SASOL (in South Africa), uses fluidized-bed reactors 46 m high, 230 cm in diameter with reaction temperature of 320—360°C and pressure of 22 bar. In the slurry-bed reactors the feed gas is bubbled through a suspension of finely divided catalyst particles. It has the advantage of good temperature control thus providing greater flexibility of reaction conditions.

Each type of reactor is better suited for certain product composition. Fixed-bed reactors, for example, produce high boiling straight-chain hydrocarbons consisting, typically, of 33% gasoline hydrocarbons (C_5—C_{11}), 17% diesel and 40% heavy paraffins and higher waxes. The gasoline fraction is of low octane value and requires further treatment (isomerization or blending) before use. Fluidized-bed reactors are the best when lighter hydrocarbons are desired. A typical product composition is 72% lower molecular weight gasoline-range hydrocarbons rich in olefins and 14% oxygenated hydrocarbons. However, the product is low in diesel. Thus two or more different reactors may be operated in parallel to provide an integrated fuel plant.

Product Selectivity

The demands on selectivity of Fischer-Tropsch reactions are ever-increasing. In the earlier days of the process the concern was to improve selectivity with respect to better gasoline grade and/or diesel oil chemicals. With the realization of reaction conditions and configurations that may make the process an industrially feasible route of converting synthesis gas to industrial intermediates, more stringent conditions are being imposed on the reaction parameters to make the process more selective.

Selectivity improvement is sought with respect to product properties such as chain length, chain branching, olefin content, alcohol content and methane content. Reaction conditions that particularly eliminate or minimize carbon deposition are desirable. In order to achieve and improve product selectivity the optimization of the following reaction parameters has been investigated: reaction temperature and pressure, H_2/CO ratio, conversion, space velocity, amount and type of promoters, nature of the catalyst, size of catalyst particles and mode of its deposition, type of support, and type of the reactor.

We have already seen examples of how the choice of the metal catalyst and support affects the product distribution. The effect of the choice of the reactor type on the nature of the reaction products has also been demonstrated.

Selectivity to olefins may be enhanced by addition of promoters such as K_2O, Ti, Mn or V. Selectivity of greater than 70% to C_2—C_4 olefins at high conversion rate has been achieved.

The search for selectivity to lower olefins by controlling the chain growth and inhibiting hydrogenation has been followed in three directions.

(a) The use of highly dispersed catalysts either by improving the method of deposition or using special dispersing supports.
(b) The use of bimetallic catalysts e.g. with Mn/Fe ratio of 9 : 1 at 330°C up to 90% olefin selectivity has been achieved. However, the activity of the catalyst and its life-time are low.
(c) The use of shape-selective catalysts. These will be discussed in conjunction with the methanol-to-gasoline conversion.

Since the Fischer-Tropsch process was originally intended for the production of hydrocarbons, little attention was paid in the early phases of the application of the process to the oxygenated products obtained as co-products. With the search for more economical sources of oxygenated hydrocarbons, the possibility of tuning the Fischer-Tropsch process for the production of such oxygenates has been investigated. It has been found that with a nitrided iron catalyst, selectivity for alcohols may exceed 80%. A $Rh/Hg/SiO_2$ catalyst system gave 75% selectivity with respect to ethanol. The major side products are olefins.

Mechanism

The Fischer-Tropsch process may be defined as being the hydrogenative oligomerization of CO over heterogeneous catalyst to produce alkanes, alkenes, oxygenated hydrocarbons and water. A complete mechanism must account for the formation of all products observed. The first attempt at eludicidating the mechanism of the process was made by the inventors of the process themselves, Fischer and Tropsch.

(1) *The surface carbide mechanism*

Fischer and Tropsch apparently were trying to explain the formation of alkanes and alkenes rather than introduce a mechanism for the whole line of products that could be formed from the process. They observed that hydrocarbon formation occurred with heterogeneous metal catalysts (Ni, Fe, Co, Ru) that tend to adsorb

CO dissociatively to form surface carbide species. Their mechanism consists of the following steps:

(i) *Initiation*:

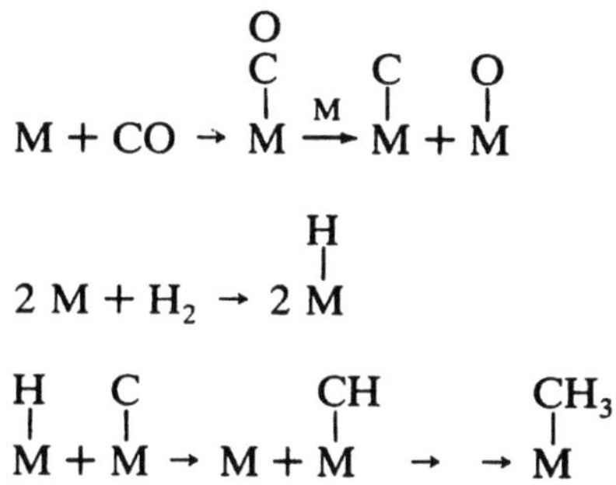

Note: M—X means species X chemisorbed on the metal surface atom M. The hyphin has no implication with respect to the strength of the M—X interaction or the order of the bond. More recently it has been suggested, based on spectroscopic evidence, that in order for an adsorbed CO to undergo dissociation it must be bonded side-on to the metal, not end-on.

(ii) *Propagation*:

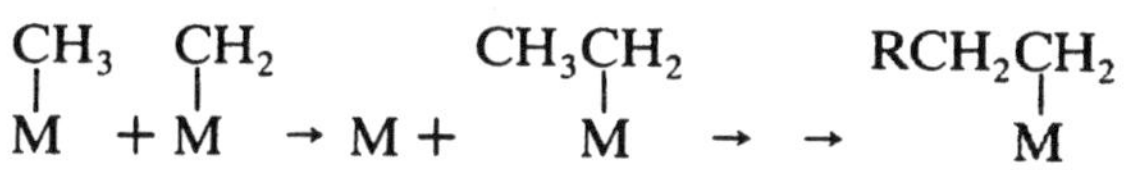

(iii) *Termination*:

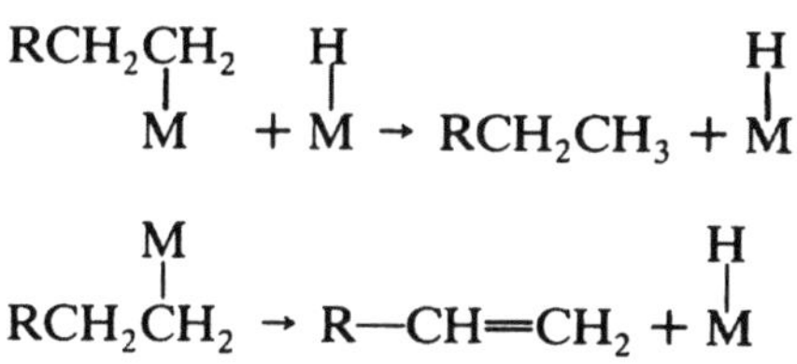

The carbide mechanism has survived the several decades since its introduction. Several items of evidence arising from recent investigations support this mechanism e.g.

(a) In the hydrogenation of CO over clean Ni surface it has been observed that CO_2 evolution precedes that of CH_4, thus

$$2\,CO \rightarrow C(ad) + CO_2$$

$$C(ad) + 2\,H_2 \rightarrow CH_4.$$

(b) Decomposition of diazomethane (CH_2N_2) at 200°C in the presence of H_2 over

Co, Fe and Ru catalysts gives linear alkanes and olefins with distribution similar to that obtained from CO/H_2 reaction over the same catalysts.

(c) Distribution of ^{13}C in $CH_2{=}CHCH_3$ formed when ^{13}CO, H_2 and $^{12}CH_2N_2$ were reacted under Fischer-Tropsch conditions is consistent with the distribution predicted based on the carbide mechanism and inconsistent with the other proposed mechanisms.

However, a drawback of this mechanism is that it does not explain the formation of oxygenated products.

(2) *The hydroxycarbene mechanism*

(i) *Initiation*:

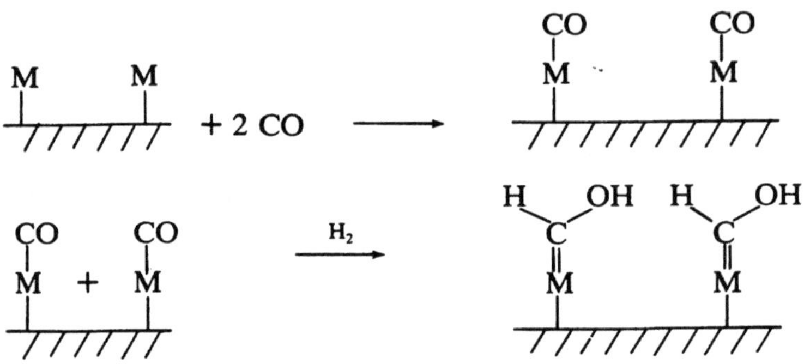

(ii) *Propagation*:

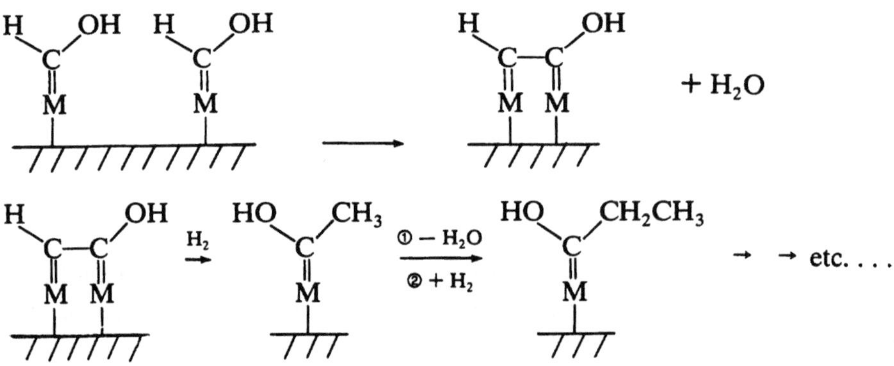

(iii) *Termination*:

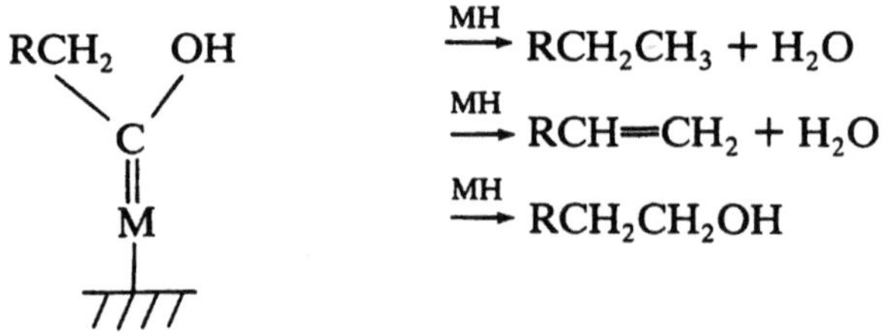

This mechanism explains the formation of alkanes, and olefins as well as oxygenated hydrocarbons. However, it precludes the dissociation of CO, which is not consistent with many experimental observations.

(3) *The CO insertion mechanism*

(i) *Initiation*:

The initiation of active species is similar to that of the carbide mechanism although the mechanism of its formation is different.

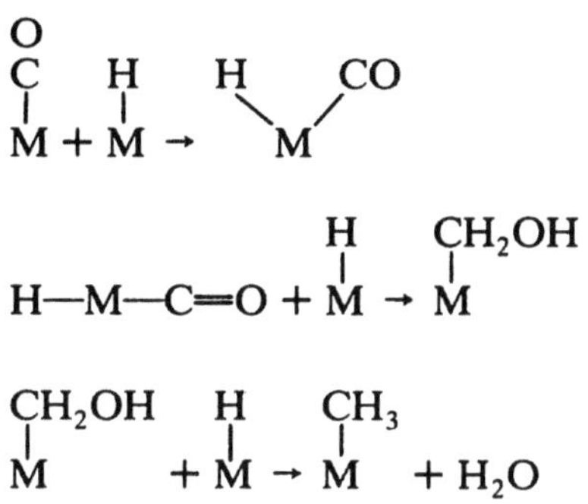

(ii) *Propagation*:

Propagation proceedes via CO insertion rather than $—CH_2—$ insertion.

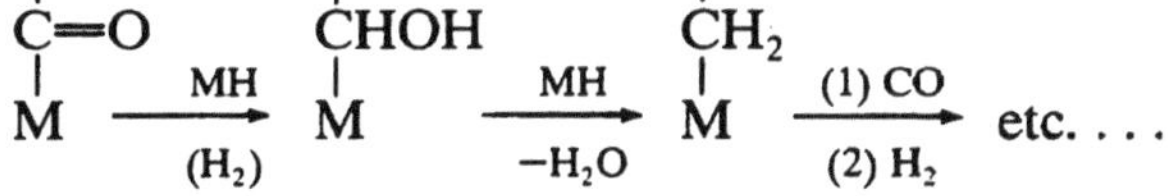

(iii) *Termination*:

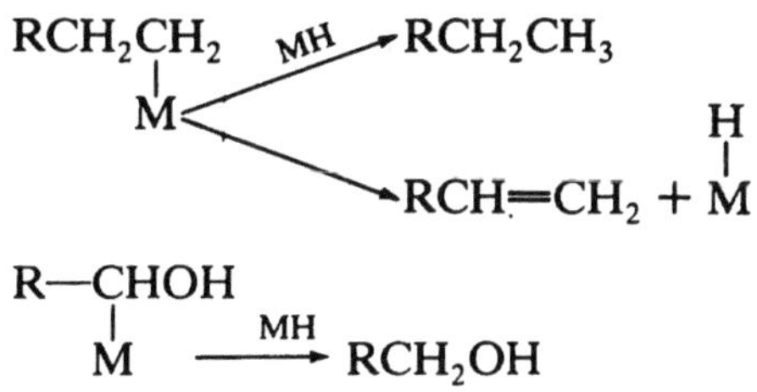

Methanol to Gasoline (MTG)

One of the most significant developments in the catalytic petrochemical processes related to the Fischer-Tropsch synthesis is the process of coverting methanol to gasoline discovered by Mobil investigators in the mid-70s. Its significance lies in the synthesis and use of shape-selective zeolite catalysts. These are crystalline aluminosilicates consisting of tetrahedral network of SiO_4 and AlO_4^- units. The anionic charges are balanced by cations such as Na^+ or H^+. These cations may also be replaced by other cations that may have some added catalytic activity to make the catalyst useful in Fischer-Tropsch type syntheses. The aluminosilicate tetrahedra enclose cavities among them that can allow through molecules of the right dimensions only. Zeolite catalysts may be synthesized (e.g. ZSM-5) that have pores of the proper size to exclude molecules with critical dimensions greater than C_{10} aromatics, which is the end point of conventional gasoline. Using this catalyst at 370°C and 15 bar converts methanol to a mixture of aliphatic and aromatic hydrocarbons in the gasoline boiling range (C_5—C_{10}), with unleaded research octane number of 90—95. The hydrocarbon mixture obtained by shape-selective catalysts is superior in quality to that obtained from the Fischer-Tropsch process (Table 6-3).

Table 6-3. Product distribution from typical Fischer-Tropsch and zeolite processes.

Product		Process	
		Fischer-Tropsch	Shape-selective (ZSM-5)
Light Gas	C_1—C_2	20.1	1.3
LPG	C_3—C_4	23.0	17.8
Gasoline	C_5—C_{12}	39.0	80.9
Diesel Oil	C_{13}—C_{18}	5.0	0
Heavy Oil	C_{19+}	6.0	0
Oxygenates		7.0	0
Aromatics (% of gasoline)		5.0	38.6

The MTG process is usually carried out in two consecutive steps. In the first step methanol is dehydrated to dimethylether. The latter is then led to the reactor containing the ZSM-5 zeolite to be converted to a mixture of hydrocarbons that are cracked under the reaction conditions, *in situ*, to gasoline components (aromatics and alkanes).

Intensive research has been carried out in attempts to adopt the MTG process to the economic conversion of methanol to lower olefins. No distinct success has been reported.

Homologation of Methanol:

Another synthetic process related to the Fischer-Tropsch synthesis, in that it converts C_1 compounds to higher hydrocarbons, is the homologation of alcohols, particularly methanol. The process consists of the reaction of methanol with synthesis gas using $Co_2(CO)_8$ as a homogeneous catalyst at 185°C and 270 bar. The reaction has 39% selectivity and 70% conversion.

$$CH_3OH + CO + 2\ H_2 \rightarrow CH_3CH_2OH + H_2O$$

More recent investigations have shown that the reaction can be promoted by iodine and its selectivity may be improved by triphenylphosphine. A system consisting of cobalt carbonyl, idoine carrier and triphenylphosphine gave 90% selectivity at 200°C and 200 bar.

The importance of methanol homologation is that, when integrated with a methanol synthesis facility (from synthesis gas), it will provide a source of ethanol essentially from synthesis gas. Ethanol in turn could easily be converted to several bulk chemicals such as acetaldehyde, butadiene etc.

The homologation reaction is of general applicability; it can be realized with many R—X compounds where R = alkyl, cyclo-alkyl or benzyl; and X = OH, ester, acid, aldehyde or ketone.

$$R{-}X \rightarrow RCH_2X$$

6.3.3. AMMONIA AND RELATED PRODUCTS

Currently, ammonia manufacture accounts for about 80% of natural gas consumption as a petrochemical feedstock. This is distantly followed by methanol (19%) and OXO alcohols (1%).

Manufacture of all synthetic ammonia is carried out by passing a combination of nitrogen and hydrogen over an iron oxide catalyst promoted with aluminum or potassium oxide, at about 450°C (430—480°C) and about 250 bar.

$$N_2 + 3\ H_2 \rightarrow 2\ NH_3$$

Nitrogen is usually obtained by feeding air to the system, but the sources of hydrogen are many and will vary considerably for each plant. In the United States, most of the hydrogen feed comes from steam reforming of natural gas. The overall ammonia plant usually consists of several main parts: steam reforming of natural gas or naphtha, carbon monoxide shifting, carbon dioxide removal, nitrogen generation, compression and conversion to ammonia.

Partial oxidation of hydrocarbons as a source of hydrogen offers flexibility in the choice of feedstock as well as ability to change feeds. It is normally used in heavier hydrocarbons including crude and residual oils, coke-over gas, and even solids, such as coal or lignite. Several companies have developed this process, which is continuous and non-catalytic under highly controlled operating conditions. Quite often an on-site air separation plant is used to supply the oxygen as well as the nitrogen.

Shift conversion of CO to CO_2 is used to remove CO and recover more hydrogen for ammonia synthesis. Carbon monoxide poisons the ammonia catalyst, and any source of hydrogen containing carbon monoxide must be treated. These objectives are accomplished in a two-step 'shift conversion'.

The first step converts most of the CO with steam at 400—480°C to CO_2 and H_2 over a chromium-promoted iron catalyst. The second step, at lower temperatures of 140—170°C, further reduces CO to under 0.5% over copper-zinc-chromium oxide catalyst. The hydrogen thus produced is also used to adjust the N_2 and H_2 ratio.

Removal of carbon dioxide produced by the shift conversion is achieved by treating the effluent gas with solvents or aqueous solutions of potassium carbonate (with and without activators) or monoethanolamine (MEA). Residual CO_2 and CO still present in trace amounts are removed by 'methanation'.

World scale ammonia plants are 1000 to 1500 tons/d in size. the potential savings per ton of NH_3 increase with plant size to this range, but begin to diminish as units of 2000 tons/d in size are approached. There is even some argument that there is no economic incentive in surpassing this size range. The economics of the production of ammonia may be improved a great deal if it is integrated with the use of reformate synthesis gas for other industries such as the synthesis of methanol. Many manufacturers use both components of synthesis gas (CO and H_2) in industries integrated in one industrial complex. Downstream utilization of ammonia will further improve the economics of its production.

Other than the production of sulfuric acid, ammonia is produced in greater volume than any other chemical. The major end use for ammonia is fertilizer applications which account for nearly three-fourths of the total consumption.

In fertilizer applications, the trend has been to direct application of anhydrous ammonia to the soil. Ammonium nitrate consumes ammonia directly as well as indirectly as a precursor for the nitric acid required for the manufacture of the nitrate. Urea is gaining wide acceptance as a solid, nitrogen-containing slow-acting fertilizer.

In chemical uses the major consumer of ammonia is acrylonitrile followed by urea and melamine, adipic acid and hexamethylenediamine.

Urea: The technical production of urea is based on the reaction of ammonia with carbon dioxide:

$$2\ NH_3 + CO_2 \rightarrow NH_2{-}COONH_4 + 126\ kJ/mol \qquad (1)$$

$$NH_2{-}COONH_4 \rightarrow (NH_2)_2CO + H_2O - 29\ kJ/mol \qquad (2)$$

The intermediate, ammonium carbamate, is formed directly from the gases. The method was developed by A. Basaroff (1870). The reaction conditions are usually 100 bar at 150—200°C. The reaction in Eq. (1) proceeds backwards above 60°C at atmospheric pressure. At elevated pressure, urea is formed and water condensed. The main problem in this reaction is to increase the yield, by recycling of the unreacted reagents. This can be achieved by many different methods.

Until 1955 the reactors had to be lined with silver, lead, or Monel metal to reduce corrosion by sulfur impurities in the carbon dioxide. Today special Cr—Ni steels have virtually solved the problem. These steels can tolerate sulfur in the presence of oxygen. The latter forms a corrosion-resistant surface oxide. Today, the main process improvements are aimed at heat recovery and yield.

Urea is an amazing chemical. It can react in three fundamentally different ways:

1. With acids, such as sulfuric acid, phosphoric acid, hydrochloric, and nitric acid, urea forms salts of the type: $NH_2 \cdot CO \cdot NH_3 \cdot NO_3$. The vibration spectra show that the ions have the structure $NH_2CONH_3^+NO_3^-$. Sodium and potassium salts are formed from the amides in liquid ammonia; $NaNH \cdot CO \cdot NH_2$.
2. With many substances, urea forms clathrates and adducts. Examples are phenol, hydroquinone, formic acid, salicyclic acid, trichloroacetic acid, ammonia and hydrogen peroxide.
3. Urea undergoes chemical reactions typical of the carbonyl-amide functional groups.

Urea-Formaldehyde (UF) Resins

Basic UF-resin chemistry consists of a stepwise addition of urea and formaldehyde to a resin of modest molecular weight. The manner of interaction is determined by the functional groups. Urea reacts as an amine, formaldehyde as a glycol. During the formation of the initial reaction products, water is released. The reaction path depends on pH, concentration, molar ratio, temperature, reaction time, and on the sequence of reactions. Obviously, these parameters permit a large number of permutations, and make for a complex system, belying the fact that only two reagents with simple functional groups are present.

Currently, a new wave of interest has drawn attention to UF reactions. This is due to progress in advanced instrumental analytical methods that make it possible to determine the composition of resins during manufacture.

References

Pearce, R. and Patterson, W. R.: *Catalysis and Chemical Processes*, John Wiley and Sons, 1981.

Sheldon, R. A.: *Chemicals from Synthesis Gas*, D. Reidel Publishing Co., 1983.

Shilov, A. E.: *Activation of Saturated Hydrocarbons by Transition Metal Complexes*, D. Reidel Publishing Co., 1984.

Keim, W.: *Catalysis in C_1 Chemistry*, D. Reidel Publishing Co., 1983.

Gates, B. C., Katzer, J. R., and Schuit, G. C. A.: *Chemistry of Catalytic Processes*, McGraw-Hill Book Co., 1979.

Rostrup-Nielsen, J. R.: 'Catalytic Steam Reforming' in *Catalysis: Science and Technology*, Edited by Anderson, J. R. and Boudart, M., Vol. 5, 1984, pp. 1—118.

Rofer-DePoorter, C. K.: 'A Comprehensive Mechanism For the Fischer-Tropsch Synthesis', *Chem. Rev.* **81**, 447—474 (1981).

Graboski, M. S.: 'The Production of Synthesis From Methane, Coal and Biomass' in Herman R. G. (Ed.) *Symposium on Catalytic Conversions of Synthesis Gas to Alcohols and Chemicals*; 1983, White Haven, Pa., Plenum Press, New York, 1984.

CHAPTER 7

Oxo and Carbonylation Reactions

7. Introduction

Many organic compounds react with carbon monoxide in the presence of a nucleophile and a suitable catalyst to give aldehydes, carboxylic acids, esters, ketones etc. The versatility of this type of reaction has led to many industrial applications. Many of these reactions have special names. The most important are:

Oxo process or hydroformylation (5 million tons per year)

$$R{-}CH{=}CH_2 + CO + H_2 \xrightarrow[\text{[Co, Rh]}]{} R{-}CH_2{-}CH_2{-}CHO \qquad (1)$$

Carbonylation of olefins or hydrocarboxylation (50,000 t/y)

$$R{-}CH{=}CH_2 + CO + H_2O \xrightarrow[\text{[Co, Ni]}]{} R{-}CH_2{-}CH_2{-}COOH \qquad (2)$$

Oxidative carbonylation of ethylene

$$CH_2{=}CH_2 + CO + \tfrac{1}{2}\,O_2 \xrightarrow[\text{[Pd, Cu]}]{} CH_2{=}CH{-}COOH \qquad (3)$$

Carbonylation of acetylene (Reppe reaction)

$$HC{\equiv}CH + CO + H_2O \xrightarrow[\text{Ni}]{} CH_2{=}CH{-}COOH \qquad (4)$$

$$2\ HC{\equiv}CH + 3\ CO + H_2O \xrightarrow[\text{Co, Ru–Rh}]{} HO{-}C_6H_4{-}OH + CO_2 \qquad (5)$$

Carbonylation of methanol

$$CH_3OH + CO \xrightarrow[\text{[Rh, Co]}]{} CH_3COOH \qquad (6)$$

Homologation of methanol

$$CH_3OH + CO + H_2 \xrightarrow[\text{[Co/Ru]}]{} CH_3{-}CH_2OH \qquad (7)$$

Homologation of formaldehyde

$$CH_2O + CO + H_2 \xrightarrow[\text{[Ru]}]{} HOCH_2{-}CHO \xrightarrow{+H_2} \underset{OH}{CH_2}{-}\underset{OH}{CH_2} \qquad (8)$$

Ethylene glycol via direct synthesis gas conversion

$$2\ CO + 3\ H_2 \xrightarrow[\text{[Rh/Ru]}]{} \underset{OH}{CH_2}{-}\underset{OH}{CH_2} \qquad (9)$$

Hydrocarbonylation of propene (Reppe)

$$CH_3{-}CH{=}CH_2 + 3\,CO + 2\,H_2O \xrightarrow[+R_3N]{Fe(CO)_5} CH_3{-}CH_2{-}CH_2{-}CH_2OH + CH_3{-}\underset{\displaystyle CH_2OH}{\underset{|}{CH}}{-}CH_3 \quad (10)$$

The carbonylation is normally catalyzed in the liquid phase by transition metal complexes. Traditionally carbonyl complexes of nickel and cobalt were used as homogeneous catalysts. Modern processes use rhodium and sometimes ruthenium complexes.

Of all these processes the Oxo reaction (equation 1) is the most important. The carbonylation of methanol with the rhodium iodide catalyst (equation 6) ranks second. It is a relatively new process, but it is expected that this process will dominate the manufacture of acetic acid in the future. The other processes are applied on a smaller scale in industry or are of only potential industrial importance. During the 1970s and early 1980s, when the crude oil prices were high, reactions such as the homologation of methanol and the glycol synthesis found considerable interest. It was thought that coal-based synthesis gas could replace oil as the major feedstock of the chemical industry. The rapid oil price decrease in the mid 1980s prevented these new syn-gas processes from going on stream.

7.1. The Oxo Process

The Oxo process, also known as hydroformylation, is the reaction of an olefin with carbon monoxide and hydrogen. It is one of the most important homogeneous catalytic processes today. About 5 million tons of Oxo products are produced every year. The reaction was discovered by O. Roelen at the Ruhrchemie in Germany (1938).

The largest single process is the hydroformylation of propene to give butanals as the primary products. Both possible isomers are formed but *n*-butanal is the more valuable product. It is either hydrogenated to *n*-butanol or transformed to 2-ethylhexanol via an aldol condensation and a subsequent hydrogenation. The alcohols are used as solvents or, more important, as plasticizers for polyvinyl chloride (PVC). For this purpose they are often converted to the corresponding phthalates by a reaction with phthalic acid anhydride.

$$CH_3{-}CH{=}CH_2 + CO + H_2 \xrightarrow{[catal.]} CH_3{-}CH_2{-}CH_2{-}CHO + CH_3{-}\underset{\displaystyle CH_3}{\underset{|}{CH}}{-}CHO \quad (11)$$

$$CH_3{-}CH_2{-}CH_2{-}CHO + H_2 \xrightarrow{[catal.]} CH_3{-}CH_2{-}CH_2{-}CH_2OH \quad (12)$$

$$2\ CH_3{-}CH_2{-}CH_2{-}CHO + H_2 \xrightarrow{[OH]} CH_3{-}CH_2{-}CH_2{-}CH_2{-}\underset{\displaystyle CH_2{-}CH_3}{CH}{-}CH_2OH \qquad (13)$$

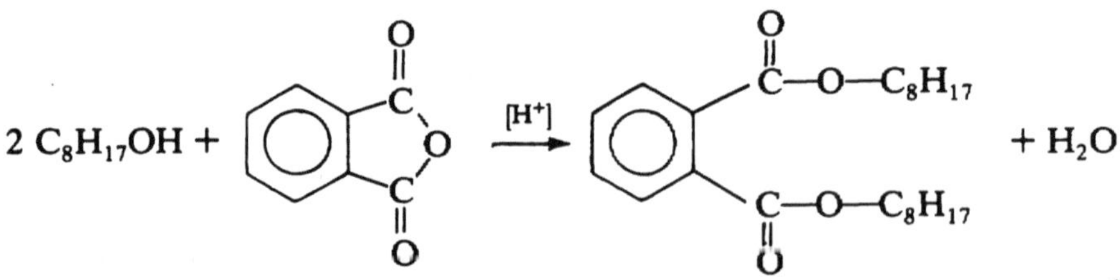

(14)

Several commercial Oxo processes are presently in operation. They use either cobalt or rhodium catalysts. Where cobalt compounds are used as the catalyst precursors, the active catalyst is hydridocarbonyl cobalt. This catalyst requires temperatures of about 150°C for reasonable reaction rates. Under these conditions a high CO partial pressure is required to stabilize the catalyst. Therefore the synthesis gas pressure is typically 30 MPa.

In the Shell process tributyl phosphine is added to the cobalt catalyst. The $HCo(CO)_3PR_3$ formed *in situ* is relatively stable. Therefore a low CO partial pressure is sufficient and the total synthesis gas pressure is only about 5 MPa. However this phosphine modified cobalt catalyst is less active than the unmodified cobalt carbonyl. Hence higher temperatures (180°C) are necessary to achieve reasonable reaction rates.

A major breakthrough came with the discovery of a rhodium catalyst by Union Carbide. This catalyst is stable, very active and has a high selectivity for linear products. It operates under much milder conditions than the cobalt-catalyzed processes. These advantages outweigh the high price of the rhodium.

Table 7-1. Commercial processes for the hydroformylation of propylene.

Process	Catalyst	Conditions	Remarks
Ruhrchemie	Co^{2+}, Co^0	150°C, 30 MPa	(a) 70% n
BASF	$HCo(CO)_4$	"	(b) "
Kuhlmann	$HCo(CO)_4$	"	(c) "
ICI	Co^{2+}	high pressure	(d) "
MAFKI	$HCo(CO)_4$	not available	(e) "
Shell	Co/PR_3	180°C, 5 MPa	(f) 88% n
UCC	$HRh(CO)(PPh_3)_3$	100°C, 3 MPa	(g) 94% n

(a) *In situ* generation of $HCo(CO)_4$; (b) $HCo(CO)_4$ is generated from Co^{2+} in a separate reactor; (c) $HCo(CO)_4$ is recycled as $NaCo(CO)_4$; (d) hydrogen-rich syngas is used; (e) direct hydrogenation to alcohols, sulfur-containing syngas can be used; (f) $HCo(CO)_3PR_3$ is generated *in situ*, low activity, by-products; (g) developed by Union Carbide/Davy Powergas/Johnson Mathey.

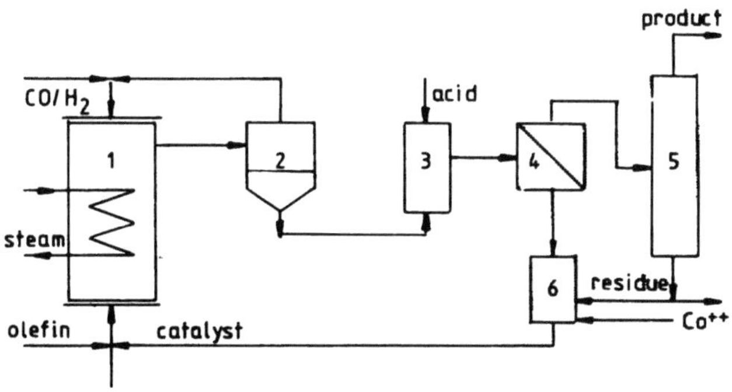

Fig. 7-1. Flow diagram of the Ruhrchemie Oxo process. 1 = Stirred tank reactor; 2 = gas—liquid separator; 3 = cobalt precipitator; 4 = cobalt separator; 5 = distillation; 6 = catalyst make-up: fresh and recycled cobalt are mixed with distillation residue.

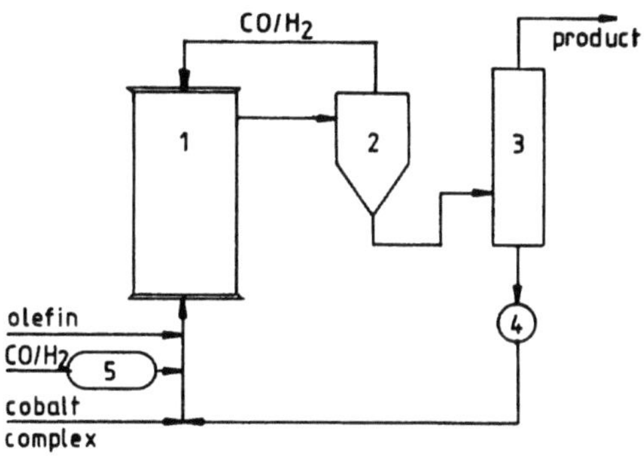

Fig. 7-2. Flow diagram of the SHELL Oxo process. 1 = Reactor; 2 = gas—liquid separator; 3 = distillation; 4 = cobalt complex separation from distillation residue; 5 = synthesis gas purification (removal of H_2S and O_2).

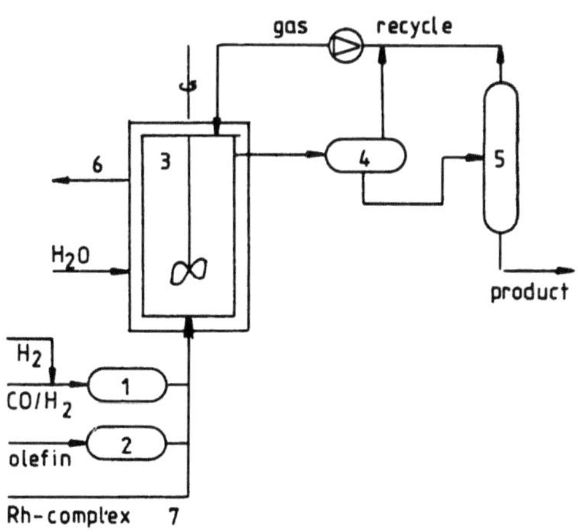

Fig. 7-3. Flow diagram of the UCC rhodium catalyzed Oxo process. 1 = Gas purification, hydrogen is added to maintain a constant partial pressure of H_2 in the reactor to stabilize the catalyst; 2 = olefin purification; 3 = stirred tank reactor; 4 = flash tank to separate synthesis gas; 5 = distillation; 6 = the reactor is cooled externally with water; 7 = the concentration of the rhodium complex is maintained at approximately 400 ppm with a 10 fold excess of triphenyl phosphine.

Schematic flow charts of industrial processes using either cobalt or rhodium catalysts are depicted in Figures 7-1, 7-2, and 7-3, and general details are given in Table 7-1.

7.1.1. THE MECHANISM OF THE COBALT CARBONYL CATALYZED HYDROFORMYLATION

Cobalt carbonyl was the first catalyst used for the hydroformylation. The mechanism has been studied by many researchers. The general reaction sequence is quite clear today. However, some details are still a matter of discussion.

The basis for all discussions was a mechanism proposed by Heck and Breslow in the early 1960s. The key steps are the formation of the alkyl complex R—Co(CO)$_4$, the acyl complex R—CO—Co(CO)$_4$, and finally the aldehyde R—CHO.

This reaction sequence was derived from stoichiometric experiments, in which equimolar amounts of the olefin and the tetracarbonylhydrido cobalt were mixed in an atmosphere of carbon monoxide.

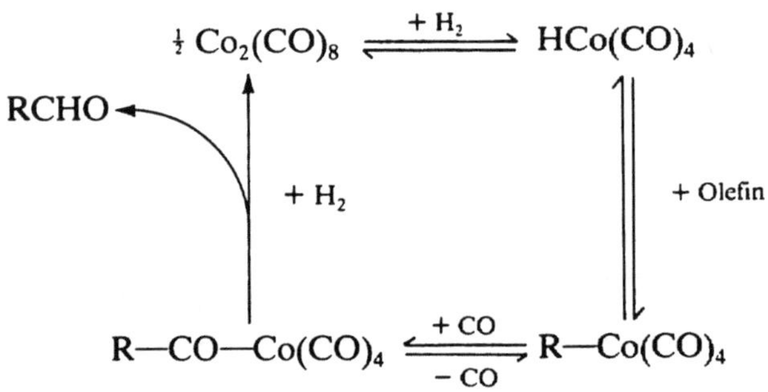

Scheme 7-1. Simplified hydroformylation mechanism according to Heck and Breslow.

As shown above the industrial cobalt-catalyzed hydroformylation proceeds at elevated temperatures and pressures. Therefore it was not clear whether the results of the stoichiometric experiments could be transferred to the high pressure conditions. However in situ IR spectroscopy with special pressure-resistant IR cuvettes fully confirmed the Heck and Breslow mechanism. Because of the importance of the Oxo process its mechanism is discussed in more detail: in the first step dicobalt octacarbonyl or other cobalt salts are converted to tetracarbonylhydrido cobalt, which is the carrier of the catalytic activity.

$$Co_2(CO)_8 + H_2 \rightleftharpoons 2\ HCo(CO)_4 \qquad (15)$$

$$2\ Co(OOC{-}R)_2 + 3\ H_2 + 8\ CO \rightleftharpoons 2\ HCo(CO)_4 + 4\ RCOOH \qquad (16)$$

The tetracarbonylhydrido cobalt is in equilibrium with the tricarbonylhydrido cobalt (equation 17). This coordinatively unsaturated complex forms a π-complex with the olefin, which in the presence of CO rearranges to give a σ-bonded alkyltetracarbonyl cobalt. In this stage the isomeric distribution of the product is decided, since both the linear and branched alkyl complexes can be formed (equations 18 and 19).

$$HCo(CO)_4 \rightleftharpoons HCo(CO)_3 + CO \qquad (17)$$

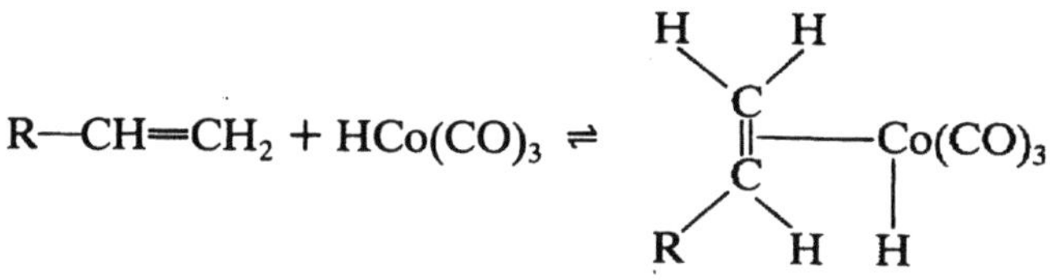

(18)

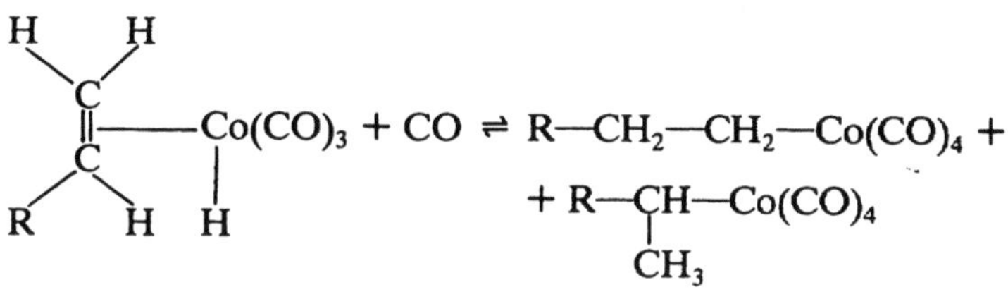

(19)

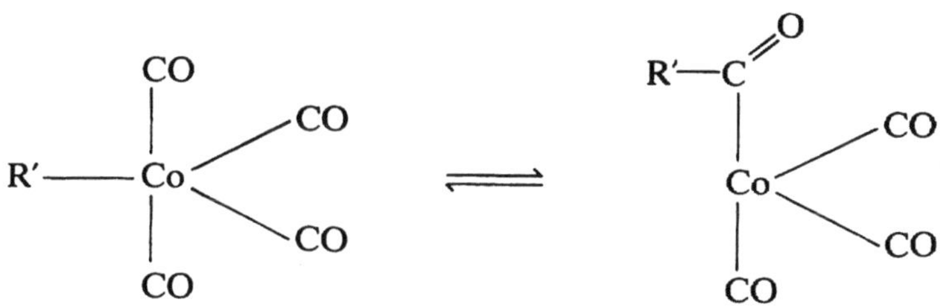

(20)

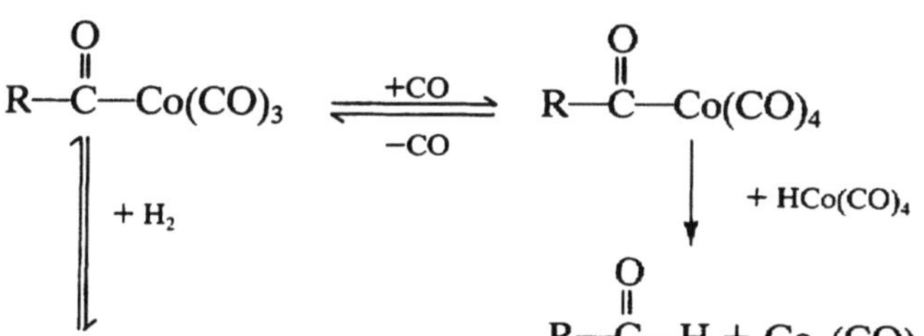

(21)

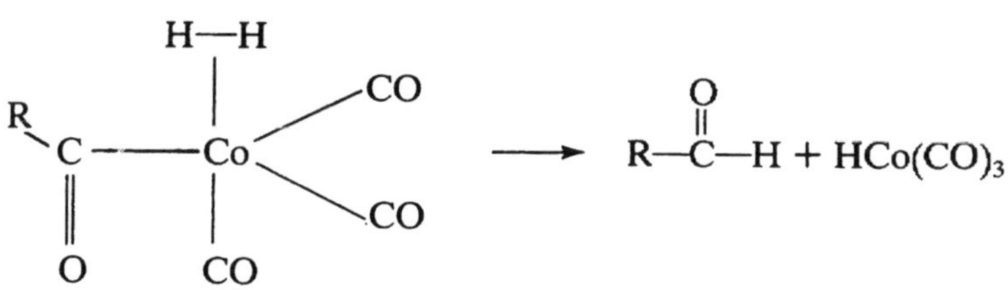

(22)

The next step is the formation of an acyl complex. It is often referred to as CO insertion. However, in reality it is an alkyl migration, as was shown by isotopic labelling. The primarily formed acyltricarbonyl cobalt can react with carbon monoxide to give a relatively stable acyltetracarbonyl cobalt or with hydrogen to give the aldehyde and tricarbonylhydrido cobalt, which can start a new catalytic cycle. It was shown in stoichiometric experiments that the acyltetracarbonyl cobalt can react with $HCo(CO)_4$ to yield the aldehyde and octacarbonyl dicobalt. However, recently, it was demonstrated that under the high pressures and temperatures of the industrial process the reaction of the acyl complex with hydrogen is predominantly responsible for the aldehyde formation.

The formation of by-products can be explained along the same lines. The hydrogenation of the olefin occurs by a reaction of the alkyl complex with hydrogen.

$$R{-}Co(CO)_4 + H_2 \rightarrow R{-}H + HCo(CO)_4 \qquad (23)$$

Another important side reaction is the hydrogenation of the aldehyde to the corresponding alcohol. In some industrial processes this side reaction is deliberately used to produce butanol from propylene via butyraldehyde, without isolation of the aldehyde.

$$R{-}C(=O)H + H_2 \xrightarrow{HCo(CO)_4)} R{-}CH_2OH \qquad (24)$$

The mechanism of the other catalysts is similar to the Heck and Breslow mechanism. The preference for the formation of linear products can be explained by a combination of electronic and steric effects. The latter can be readily visualized by looking at the transition from the π-complex to the alkyl complex, when a phosphine ligand is present in the complex.

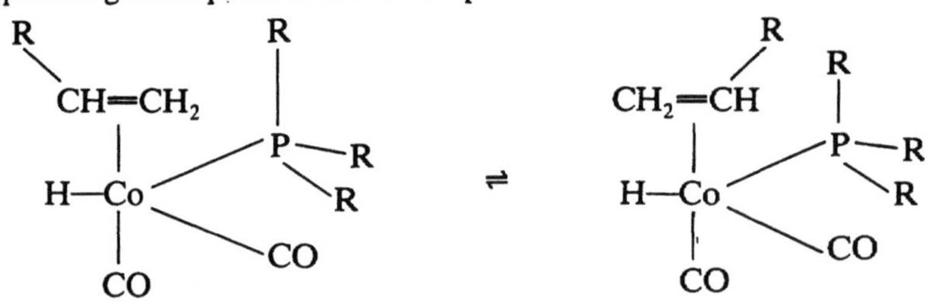

(25)

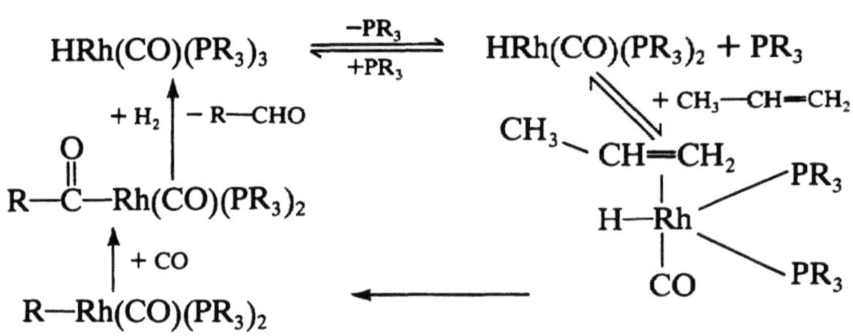

Scheme 7-2. Mechanism of the rhodium-catalyzed hydroformylation of propene.

7.2. The Carbonylation of Olefins

Olefins react with CO in the presence of a nucleophilic reaction partner of the type HX to give carboxylic acid derivatives. This class of reactions is named after their inventor as Reppe reactions. Catalysts are metal carbonyls or CO-containing complexes of noble metals.

$$CH_2{=}CH_2 + CO + HX \xrightarrow[\text{[Catal.]}]{} CH_3{-}CH_2{-}C(=O){-}X \qquad (26)$$

$$X = {-}OH,\ {-}OR,\ {-}SR,\ {-}NHR,\ {-}O{-}C(=O){-}R \text{ etc.}$$

$$\text{[Catal.]} = Ni(CO)_4,\ Fe(CO)_5,\ Co_2(CO)_8,\ \text{Rh-, Ru-, Pd-complexes)}$$

If HX represents water the reaction is called hydrocarboxylation and the product is propionic acid. This reaction takes place at 240 bar and 300°C and the catalysts are tetracarbonyl nickel or cobalt carbonyls. The yield of acid is 95% with respect to ethylene. Propionic acid is used as a starting material for esters or as a food additive.

Other olefins besides ethylene react similarly. However, a variety of isomers is formed, thereby limiting the usefulness of this reaction to special applications.

The mechanism of the Reppe reaction is similar to that of the hydroformylation. Only the catalytic cycle is initiated differently. For instance the hydridotetracarbonyl cobalt acts as a relatively strong acid in water or methanol and is therefore largely dissociated. Therefore it was proposed that the activation of the olefin occurs by an ionic addition.

$$HCo(CO)_4(aq) \rightleftharpoons H^+_{(aq)} + Co(CO)^-_{4(aq)} \qquad (27)$$

$$R'{-}CH{=}CH_2 + H^+ \rightarrow R'{-}{}^{\oplus}CH{-}CH_3 \xrightarrow{+\,Co(CO)_4^-} R{-}CH(Co(CO)_4){-}CH_3 \qquad (28)$$

$$\xrightarrow{-CO} H{-}Co(CO)_3(\leftarrow R{-}CH{=}CH_2) \rightleftharpoons R{-}CH_2{-}CH_2{-}Co(CO)_3 \qquad (29)$$

The alkyl complex then undergoes the same cycle as was proposed by Heck and Breslow for the hydroformylation.

A very similar reaction is the 'Koch'-process. Here the reaction of olefins with CO and water is catalyzed by protonic catalysts. This is a typical addition reaction following the rules of carbocation chemistry and yields only branched products. (This reaction is also discussed in Chapter 8.)

7.3. Carbonylation of Methanol to Acetic Acid

The carbonylation of methanol to acetic acid becomes economical when inexpensive synthetic methanol is available in large quantities. The traditional process was developed mainly by the BASF. It uses cobalt iodide as a catalyst and requires 250°C and 68 MPa carbon monoxide pressure.

The selectivities to acetic acid are 90% with respect to methanol and 70% with respect to CO. In the 1960s Monsanto developed a methanol carbonylation process which uses rhodium iodide as the catalyst. It is not only more selective for acetic acid but also operates under much milder conditions than the cobalt process. The complete recovery of the rhodium is essential for the economics of the Monsanto process.

$$CH_3OH + CO \xrightarrow[\text{[Catal.]}]{} CH_3COOH \qquad (30)$$

	BASF-process	Monsanto-process
Catal.:	CoI_2	RhI_2
Temp.	250°C	150°C
Pressure	68 MPa	0.3 MPa
Selectivity	90% (MeOH)	99% (MeOH)
	70% (CO)	90% (CO)

Texaco recently announced a new acetic acid synthesis directly from syngas with 80% yield. The catalyst is a melt of ruthenium and cobalt halides dispersed in tetrabutylphosphonium bromide. The catalyst precursors could be a mixture of $Ru_3(CO)_{12}$, CoI_2 and $(C_4H_9)_4PBr$. It is believed that the ruthenium catalyzes the methanol formation and that methanol is transformed by the cobalt catalyst to acetic acid via methyl iodide as the intermediate.

$$CO + \tfrac{1}{2} H_2 \xrightarrow{Ru} Ru{-}\overset{\overset{\displaystyle O}{\|}}{C}{-}H \xrightarrow{+H_2} Ru{-}CH_2OH$$

$$\xrightarrow[-Ru]{+H_2} CH_3OH \qquad (31)$$

$$CH_3OH + HI \rightleftharpoons CH_3I + H_2O \qquad (32)$$

$$CH_3I + Co(CO)_4^- \rightleftharpoons CH_3Co(CO)_4 + I^- \qquad (33)$$

$$CH_3Co(CO)_4 + CO \rightleftharpoons CH_3{-}\overset{\overset{\displaystyle O}{\|}}{C}{-}Co(CO)_4 \qquad (34)$$

$$CH_3{-}\overset{\overset{\displaystyle O}{\|}}{C}{-}Co(CO)_4 + H_2 \rightleftharpoons CH_3COOH + HCo(CO)_4 \qquad (35)$$

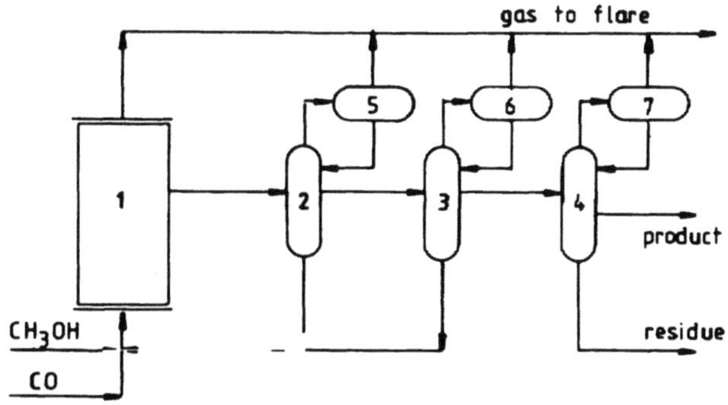

Fig. 7-4. Flow diagram of the Monsanto acetic acid process. 1 = Reactor; 2 = methanol separator; 3 = water separator; 4 = product distillation; 5, 6, 7 = gas—liquid separators.

References

General

Weissermel, K. and Arpe, H. J.: *Industrial Organic Chemistry*, Verlag Chemie, Weinheim—New York 1978.
Falbe, J.: *Carbon Monoxide in Organic Synthesis*, Springer, Berlin—New York, 1970.
Falbe, J.: *New Syntheses with Carbon Monoxide*, Springer, Berlin—New York, 1980.
Wender, I. and Pino, P.: *Organic Syntheses via Metal Carbonyls*, Vol. 2, Wiley Interscience, New York, 1977.
Graziani, M. and Giongo, M.: *Fundamental Research in Homogeneous Catalysis*, Vol. 4, Plenum New York, 1984.

Original

Roelen, O. (Ruhrchemie AG): German Patent 849 548 (1938).
Heck, R. F. and Breslow, D. S.: *J. Am. Chem. Soc.* **83**, 4024 (1961).
Mirbach, M. F.: *J. Organometal. Chem.* **265**, 205 (1984).
Slaugh, L. H. and Mullineaux, R. D.: *J. Organometal, Chem.* **13**, 469 (1968).
Pruett, R. L. and Smith, J. A.: *J. Organometal Chem.* **34**, 327 (1969).
Mirbach, M. F. and Mirbach, M. J.: *J. Mol. Catal.* **32**, 59 (1985).
Reppe, W.: *New Developments in the Chemistry of Acetylene and Carbon Monoxide*, (in German) Springer-Verlag Berlin (1949).
Roth, J. F. and Craddock, J. H.: *Chem. Technol.*, 600 (1971).
Hohenschutz, H., Kutepow, N., and von Himmele, W.: *Hydrocarbon Processing* **45**, 141 (1966).

CHAPTER 8

Acid or Base Catalyzed Addition, Elimination and Substitution Reactions

8.1. Introduction

Addition, elimination and substitution reactions are well-known in organic chemistry. Therefore it is not surprising that they have also found many applications in industrial chemistry.

$$R{-}CH{=}CH{-}R' + HX \underset{\text{Elimination}}{\overset{\text{Addition}}{\rightleftharpoons}} R{-}\underset{\displaystyle X}{\underset{|}{C}}H{-}CH_2{-}R' \tag{1}$$

$$R{-}X + HY \overset{\text{Substitution}}{\rightleftharpoons} R{-}Y + HX \tag{2}$$

Polar or polarizable compounds can be added to double bonds of olefins, ketones, nitriles etc. or to ring compounds to yield, for example, organohalogens, alcohols, mercaptans. The reverse process is called elimination. It is also widely used to produce compounds such as olefins and epoxides. Substitution reactions are those in which one functional group is replaced by another.

These three processes share common ground, in that they are catalyzed by acids or bases, whereby the catalyst can be homogeneous or heterogeneous. Homogeneous acids and bases are the well-known hydrochloric or sulfuric acids and sodium hydroxide, calcium hydroxide or ammonia. Examples of heterogeneous acids and bases are silica, alumina, metal oxides, zeolites, ion exchange resins. Sometimes special catalyst combinations are used to achieve special selectivity effects.

The homogeneous catalysts such as sulfuric acid and calcium hydroxide have a long tradition. However the acids can be recovered after the reaction only in diluted form. They must be reconcentrated in energy-consuming distillation processes or they must be disposed of by neutralization, deep-well injection or dumping into aquatic systems. Of course the latter alternative is opposed by environmentalists. A similar problem is encountered with bases, which often end up as salts, without any further industrial use. Therefore modern processes use heterogeneous catalysts which avoid these problems.

8.2. Mechanistic Considerations. Addition Reactions

Addition reactions proceed normally via an initial addition of a positive particle, often a proton, to give a positive carbonium ion.

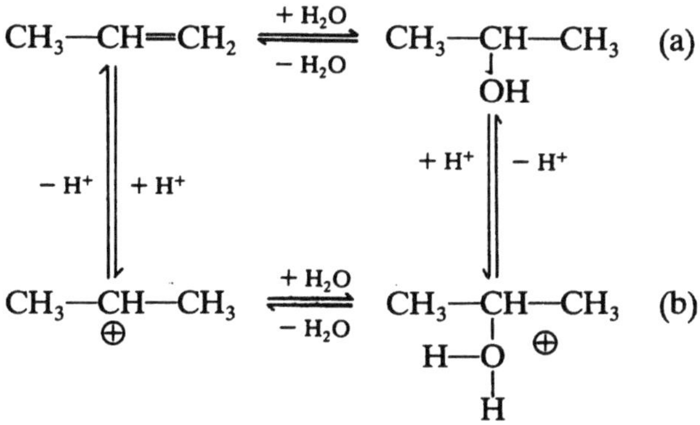
(3)

Since carbonium ions are the more stable the more branched they are, secondary or tertiary derivatives are formed whenever possible. The formation of the most branched carbonium ion is also known as the 'Markownikoff rule'. The example in equation 8.3 illustrates why isopropanol is the product of the addition of water to propene. Actually, in the case of 1-butene tertiary products can be formed as a consequence of a carbonium ion rearrangement. This is illustrated in the example of the 'Koch synthesis' of branched carboxylic acids from butene, carbon monoxide and water (equation 8.4).

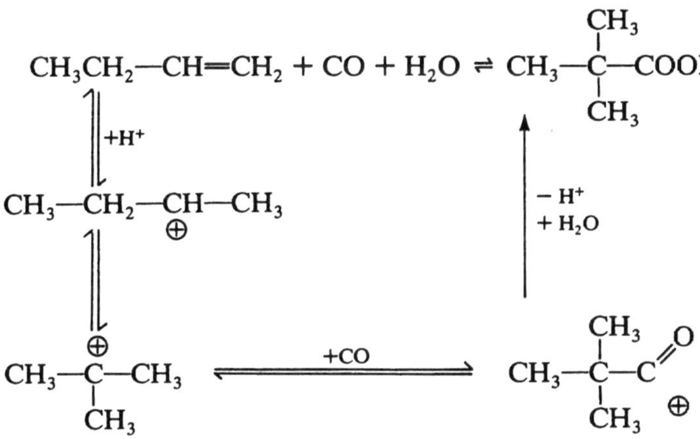
(4)

For heterogeneous acids additional adsorption and desorption steps must be taken into account.

$$H_2O(g) \rightleftharpoons H_2O(ads) \rightleftharpoons H^+(ads) + OH^-(ads) \quad (5)$$

$$C_2H_4(ads) + H^+(ads) \rightleftharpoons C_2H_5^+(ads) \quad (6)$$

$$C_2H_5^+(ads) + H_2O(g) \rightleftharpoons C_2H_5OH(ads) + H^+(ads) \quad (7)$$

$$C_2H_5OH(ads) \rightleftharpoons C_2H_5OH(g) \quad (8)$$

Often the reaction of the adsorbed olefin with an adsorbed proton is rate determining. The simplest way to produce a heterogeneous acid catalyst is to adsorb an acid on a carrier. The most common example is the combination of phosphoric acid and silica.

However, many oxides exhibit acid-base properties at elevated temperatures by themselves. Silica, for example, carries on its surface hydroxyl groups, which are acids.

$$\text{—Si—OH} + H_2O \rightleftharpoons \text{—Si—O}^- + H_3O^+ \quad (9)$$

$$\text{—Si—OH} + \text{ROH} \rightleftharpoons \text{—Si—O—R} + H_2O \quad (10)$$

Addition of metal oxides or metal salts to the silica can enhance the acidity. This is especially true for special mixed oxides such as zeolites. Synthetic ion exchangers are usually polystyrenes, which are modified by sulfonation. The sulfonic acid groups carry the catalytic activity.

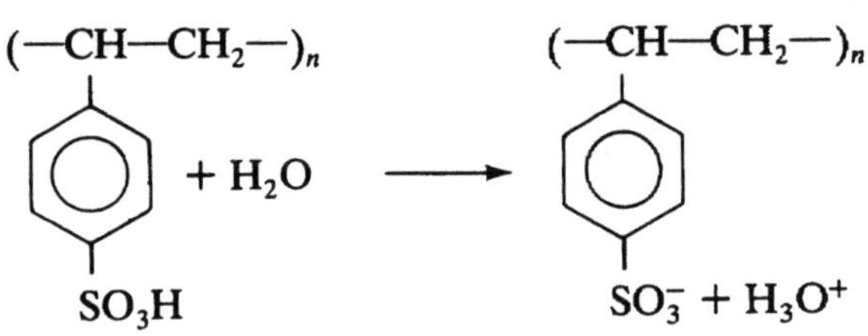

(11)

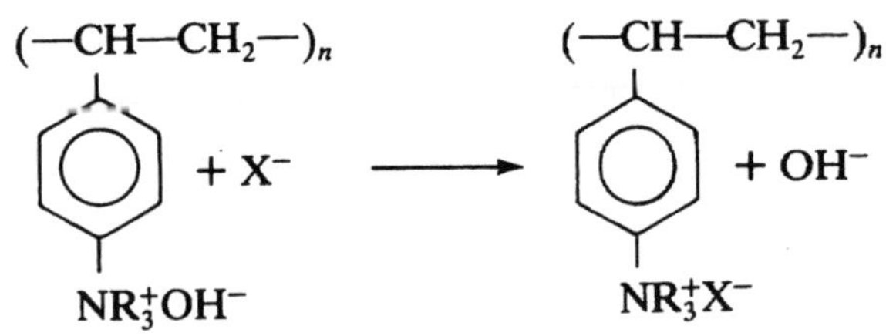

(12)

Similarly quarternary ammonium groups on a polystyrene resin can be used as basic catalysts.

The properties of the ion exchangers can be modified by crosslinking the polystyrene with divinyl benzene before sulfonation. Nevertheless the application of the resins is limited by their low thermal stability ($<150°C$) and the effect of swelling, which occurs when they are in contact with organic solvents.

8.3. Hydration of Olefins

8.3.1. ETHANOL PRODUCTION

Today most synthetic ethanol is produced by hydration of ethylene. Originally aqueous sulfuric acid was used as the catalyst. Modern processes apply heterogeneous acids as catalysts.

$$CH_2{=}CH_2 + H_2O \xrightleftharpoons{[H^+]} CH_3{-}CH_2{-}OH \qquad \Delta H = -46 \text{ kJ/mol} \qquad (13)$$

Thermodynamically the ethanol formation is favored by high pressure and low temperature. However, to obtain reasonable reaction rates, elevated temperatures are necessary. The first commercial process which used a heterogeneous catalyst was introduced by Shell in 1947. It uses phosphoric acid on silica as the catalyst and operates at a temperature of approximately 300°C and a pressure of 7 MPa. The molar ratio of water to ethylene is less than one in order to avoid catalyst losses. Under these conditions the conversion of ethylene is only about 4%. Therefore ethylene must be recycled many times. The selectivity for ethanol is about 97%.

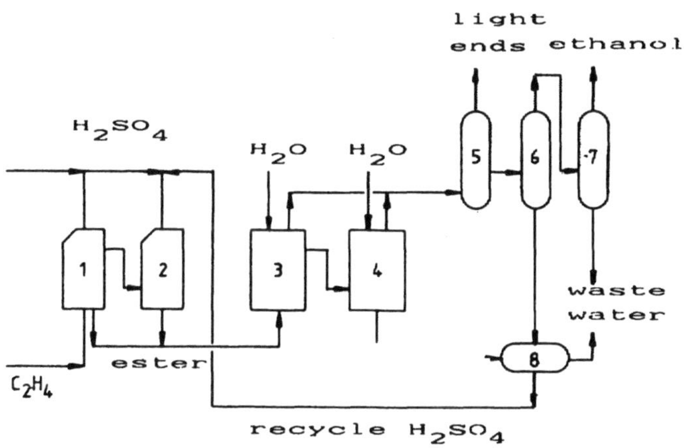

Fig. 8-1. Hydration of ethylene with sulfuric acid as a homogeneous catalyst. 1, 2 = reactors for the addition of sulfuric acid; 3, 4 = hydrolysis reactors; 5, 6, 7 = distillation columns; 8 = concentration of spent sulfuric acid.

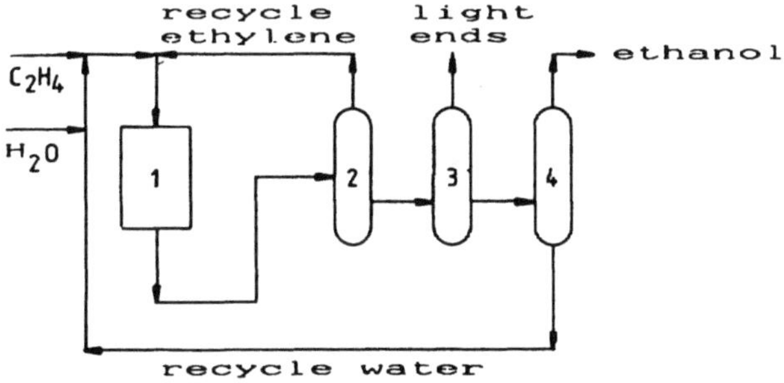

Fig. 8-2. Hydration of ethylene with a heterogeneous acidic catalyst. 1 = hydration reactor; 2, 3, 4 = distillation columns.

Other solid catalysts for the hydration of ethylene are sulfuric or boric acid supported by a porous carrier, heteropoly acids, zeolites and synthetic ion exchanger resins. Normally the activity of the catalyst is directly proportional to its acidity.

The older H_2SO_4 process is actually a multi-step reaction. First sulfuric acid reacts with ethylene to give monoethyl and diethyl sulfate, which are hydrolyzed in a second step to give ethanol and dilute sulfuric acid. The acid is concentrated in a third step and recycled.

$$CH_2{=}CH_2 + H_2SO_4 \rightarrow CH_3{-}CH_2{-}O{-}SO_3H \tag{14}$$

$$2\,CH_2{=}CH_2 + H_2SO_4 \rightarrow (CH_3{-}CH_2{-}O{-})_2SO_2 \tag{15}$$

$$CH_3{-}CH_2{-}O{-}SO_3H + H_2O \rightarrow CH_3{-}CH_2{-}OH + H_2SO_4 \tag{16}$$

$$(CH_3{-}CH_2{-}O{-})_2SO_2 + 2\,H_2O \rightarrow 2\,CH_3{-}CH_2{-}OH + H_2SO_4 \tag{17}$$

The selectivity to ethanol is approximately 86%, with diethyl ether as the major by-product.

The world capacity for synthetic ethanol is about 2 million tons per year. Ethanol is also produced by fermentation of carbohydrates in a non-catalytic process.

8.3.2. ISOPROPANOL PRODUCTION

Isopropanol is prepared by hydration of propene, in processes similar to those described for ethanol.

$$CH_3{-}CH{=}CH_2 + H_2O \xrightleftharpoons{[H^+]} CH_3{-}\underset{\displaystyle OH}{\underset{|}{C}H}{-}CH_3 \tag{18}$$

Propene is more reactive than ethylene, since the secondary propyl carbonium ion formed in the first step by addition of a proton to the olefin is more stable than the ethyl carbonium ion. The gas-phase hydration is carried out at temperatures above 200°C and at elevated pressure. The ICI process employs WO_3 on silica as the catalyst with ZnO as the promoter (27°C, 25 MPa). The Veba process uses phosphoric acid on silica as the catalyst and operates at 180—260°C and 2.5—6.5 MPa.

The hydration can also be carried out with liquid water and gaseous propene on an ion exchanger catalyst in a trickle bed reactor (Texaco-process, approx. 150°C, 8—10 MPa).

As in the case of ethylene, there is also the traditional process with aqueous sulfuric acid as the catalyst. The mechanism is analogous to that described for ethylene. The selectivity is lower than in the modern gas-phase process.

8.3.3. HIGHER ALCOHOLS BY HYDRATION OF OLEFINS

The acid catalyzed hydration of higher olefins always yields branched alcohols due to the carbenium ion rearrangement. Therefore linear alcohols are produced by other processes, such as hydroformylation, aldol condensation of acetaldehyde or the Reppe process. However *sec*- and *tert*-butanols can be manufactured by hydration of *n*- and *iso*-butenes, respectively

$$\begin{array}{l} CH_3{-}CH_2{-}CH{=}CH_2 + H_2O \xrightarrow[{[H_2SO_4]}]{} CH_3{-}CH_2{-}\underset{\displaystyle OH}{\underset{|}{CH}}{-}CH_3 \\ CH_3{-}CH{=}CH{-}CH_3 + H_2O \xrightarrow[{[H_2SO_4]}]{} \nearrow \end{array} \qquad (19)$$

$$CH_3{-}\underset{\displaystyle CH_3}{\underset{|}{C}}{=}CH_2 + H_2O \xrightarrow[{[H_2SO_4]}]{} CH_3{-}\overset{\displaystyle CH_3}{\overset{|}{\underset{\displaystyle CH_3}{\underset{|}{C}}}}{-}OH \qquad (20)$$

The direct hydration of olefins higher than propene is not commercially significant.

However a similar process, the addition of methanol to isobutene, is rapidly gaining importance. The product is methyl-tertiary butyl ether (MTBE). MTBE is used as a gasoline additive, since it has a high octane number and can replace lead alkyls as an octane booster.

$$CH_3\underset{\displaystyle CH_3}{\underset{|}{C}}{=}CH_2 + HOCH_3 \xrightarrow{[H^+]} CH_3{-}\overset{\displaystyle CH_3}{\overset{|}{\underset{\displaystyle CH_3}{\underset{|}{C}}}}{-}O{-}CH_3 \qquad (21)$$

The reaction is usually carried out in the gas phase with solid acids as catalysts, e.g. ion exchangers.

8.4. Hydration of Ethylene Oxide

Originally ethylene oxide itself was producted by an addition reaction, the chlorohydrin process. This is a two-step process. The first step is the addition of chlorine and water to ethylene. Dehydrochlorination with lime then yields the ethylene oxide.

$$Cl_2 + H_2O \rightarrow HOCl + HCl \qquad (22)$$

$$CH_2{=}CH_2 + HOCl \rightarrow HO{-}CH_2{-}CH_2{-}Cl \qquad (23)$$

$$2\,HO{-}CH_2{-}CH_2{-}Cl + Ca(OH)_2 \rightarrow \underset{\diagdown O \diagup}{CH_2{-}CH_2} + CaCl_2 + 2\,H_2O \qquad (24)$$

Since all the expensive chlorine is lost as calcium chloride, this process has been replaced today by a direct oxidation process (see Chapter 5).

However, ethylene oxide undergoes numerous addition reactions, e.g. with water, ammonia, amines, alcohols. Many of these industrial addition processes are catalytic, although the ethylene oxide also undergoes non-catalytic addition reactions.

The most important use of ethylene oxide is in the production of ethylene glycol. The catalytic hydration proceeds at normal pressure and 50—70 °C. A large excess of water is required to prevent the formation of di- and polyethylene glycols.

Glycols are used as antifreeze agents and for the production of polyesters.

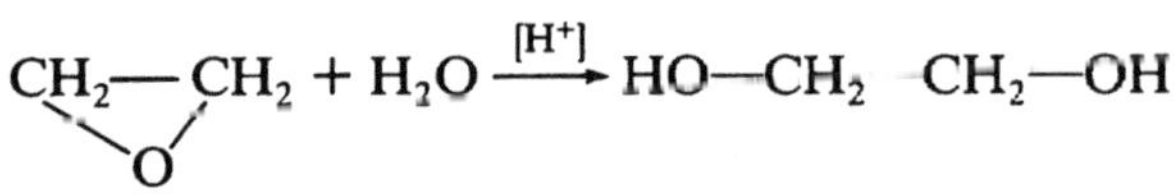
(25)

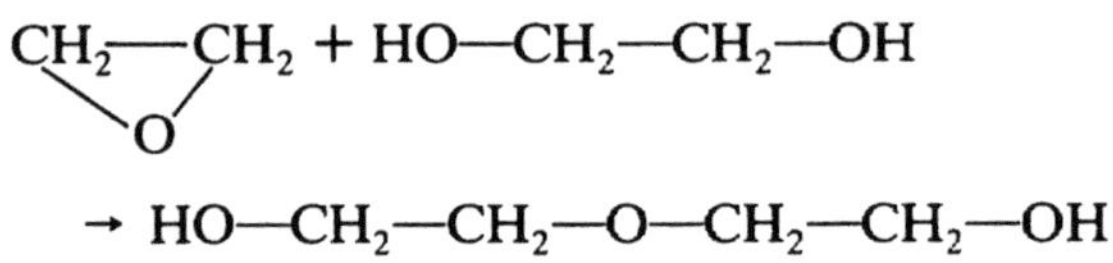
(26)

With alcohols, instead of water, ethylene oxide forms glycol ethers, which find a variety of applications as solvents and additives. The addition is similar to the hydration. However in industrial practice bases are used as catalysts.

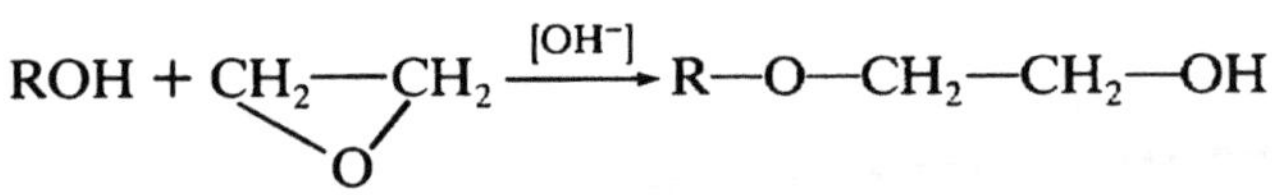
(27)

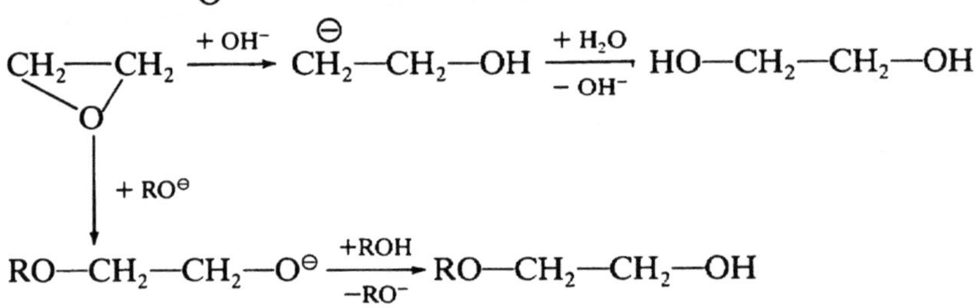
(28)

(28)

Ethylene oxide reacts with aqueous ammonia to ethanol amines at elevated temperatures and pressures. A separate catalyst is not required, since the ammonia itself is a base.

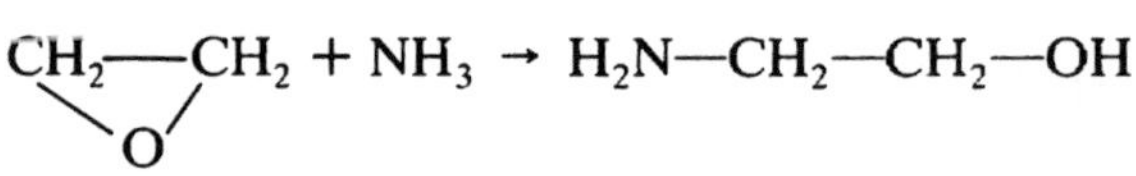
(29)

The primary ethanol amines produced in the first step can react further to give di- and triethanolamines. Ethanolamines are used for instance in gas desulfurization processes and as an intermediate in organic synthesis.

8.5. Propene Oxide and Higher Epoxides

Approximately 2.5 million tons of propene oxide are produced annually, more than half of it by the chlorohydrin process. Another 40% are manufactured via indirect oxidation processes. The direct catalytic epoxidation of propene has not so far been possible.

As in the case of ethylene the first step of the chlorohydrin process is the addition of HOCl to the propene. Two isomers, α- and β-chlorohydrin, are formed. In the second step calcium hydroxide is used to eliminate HCl. The epoxide is continuously distilled out of the reaction mixture to avoid uncontrolled hydration.

In the indirect oxidation process a peroxide or a peracid is added to propene in the presence of a catalyst to give the propene oxide.

The peroxides are produced by oxidation of hydrocarbons, isopropanol and so on with oxygen. Peracetic acid is produced by autoxidation of acetaldehyde or by treatment of acetic acid with hydrogen peroxide. The latter reaction is an acid-catalyzed substitution whereas the other oxidations proceed via a free radical pathway. They are also treated in Chapter 5.

$$RH + O_2 \rightarrow ROOH \qquad (30)$$

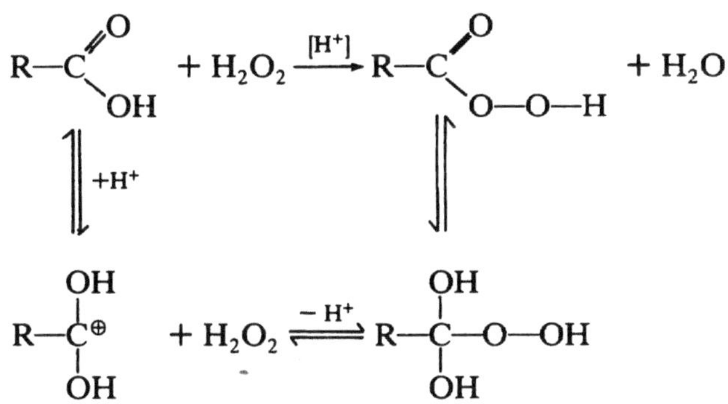

(31)

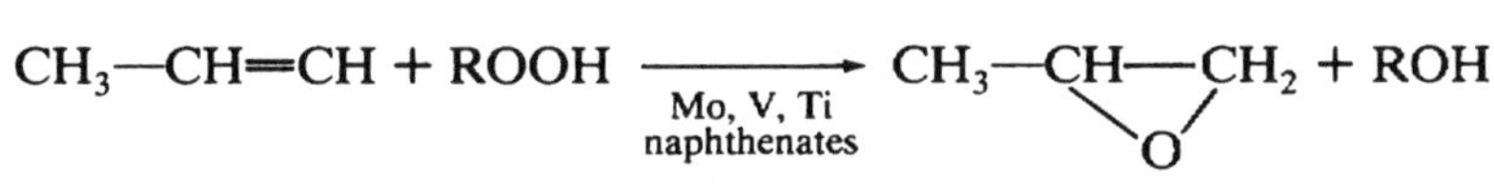

(32)

About 60% of the propene oxide is converted to polypropylene glycol or other propoxylated products by an acid- or base-catalyzed addition of propylene glycol or other polyfunctional alcohols, respectively. The propoxylated alcohols are used for the production of polyurethanes. Polypropylene glycol is nearly insoluble in water. However, by copolymerizing propene oxide and ethylene oxide the properties of the mixed polyglycols are modified over a wide range. They are used mainly as non-ionic surfactants in detergents and emulsifiers.

Thirty percent of the propene oxide is hydrated to give propylene glycol, which is used as antifreeze agent, brake fluid and in alkyd and polyester resins. The hydration is carried out with a large excess of water in the presence of acid or base catalysts. The mechanisms are analogous to those described for ethylene oxide.

8.6. Dimerization and Oligomerization of Olefins

Branched olefins are produced by dimerization or oligomerization of propene or isobutene. This reaction is formally an addition of one olefin molecule to a second. Heterogenized mineral acids, ion exchangers or organo aluminum compounds are used as catalysts.

$$CH_3{-}CH{=}CH_2 + H^+ \rightarrow CH_3{-}C^{\oplus}H{-}CH_3 \qquad (33)$$

$$CH_3{-}C^{\oplus}H{-}CH_3 + CH_3{-}CH{=}CH_2 \rightarrow \begin{array}{l} CH_3 \\ \quad \diagdown CH{-}CH_2{-}C^{\oplus}H{-}CH_3 \\ CH_3 \diagup \end{array} \qquad (34)$$

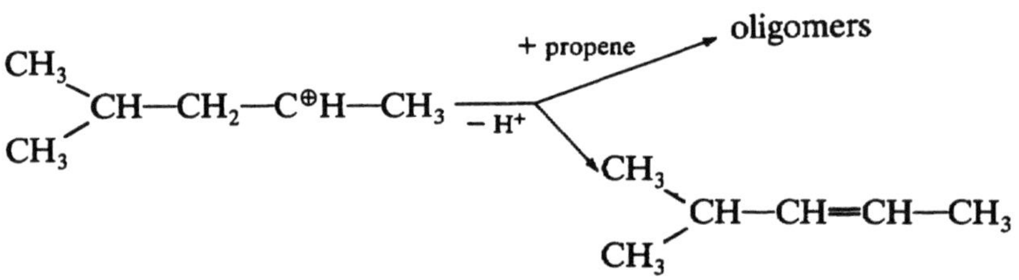

(35)

The Bayer process uses an acidic ion exchanger catalyst at 100°C and 2 MPa to dimerize and trimerize isobutene. Since isobutene reacts much faster than *n*-butene the process is suitable for removing isobutene from a C_4-olefin stream, from which nearly pure *n*-butene can be recovered for further processing. The C_8 and C_{12} olefins formed are used as an alkylating agent or after hydrogenation as a gasoline additive.

A second process was developed by UOP for the oligomerization of propene. Phosphoric acid on silica is the catalyst, which works at approximately 200°C and 50 bar pressure. The products are mainly C_9 and C_{12} branched olefins. These olefins are starting materials for the manufacture of alkylbenzene sulfonates (ABS). Since branched ABS are only slowly biodegraded, their importance as the active component in detergents has decreased due to environmental concerns.

Organo-aluminum catalysts oligomerize olefins by insertion reactions rather than by addition. Therefore they are discussed in Chapter 9.

8.7. Vinyl Compounds

8.7.1. VINYL CHLORIDE

Vinyl chloride is the monomer for the production of polyvinylchloride (PVC).

About 8 to 10 million tons are produced annually world-wide. Most of it is produced by oxychlorination of ethylene. The overall reaction is

$$4\ CH_2{=}CH_2 + 2\ Cl_2 + O_2 \rightarrow 4\ CH_2{=}CH{-}Cl + 2\ H_2O \qquad (36)$$

However, this equation does not describe the real sequence of classical addition and elimination reactions.

The first step is the addition of chlorine to ethylene with $FeCl_3$, $CuCl_2$ or $SbCl_3$ as catalysts at 40—70°C and 5 bar. According to the classical mechanism the catalyst polarizes the chlorine. The positively charged chlorine attacks the olefin to form a carbenium ion intermediate.

$$Cl_2 + FeCl_3 \rightleftharpoons Cl^+ + FeCl_4^- \qquad (37)$$

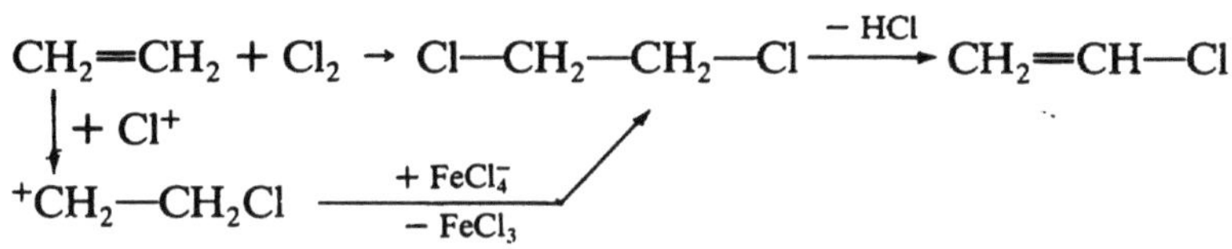

(38)

Hydrochloric acid is eliminated in the vinyl chloride forming step. It can be used as the chlorine source in the presence of oxygen and copper (II) chloride. The $CuCl_2$ is the actual chlorinating agent. It is reduced

$$CH_2{=}CH_2 + 2\ CuCl_2 \rightarrow Cl{-}CH_2{-}CH_2Cl + 2\ CuCl \qquad (39)$$

$$2\ CuCl + \tfrac{1}{2}\ O_2 \rightarrow CuO \cdot CuCl_2 \qquad (40)$$

$$CuO \cdot CuCl_2 + 2\ HCl \rightarrow 2\ CuCl_2 + H_2O \qquad (41)$$

to Cu(I) chloride, which is then reoxidized by oxygen. No free chlorine is involved. The process operates typically at 230°C and 0.2—0.4 MPa. The reaction is exothermic and the processes used by different manufacturers usually vary in the reactor design and the methods to control the evolution of heat.

In modern processes the dehydrochlorination of dichloroethane to vinyl chloride is achieved by thermal cracking in a purely thermal radical chain process at 500—600 °C and 2.5—3.5 MPa. Catalytic dehydrochlorination on pumice or charcoal at 430—530 °C and 2.5 MPa is less important.

The classical conversion of dichloroethane to vinyl chloride with alkali hydroxides in the liquid phase is rarely used today.

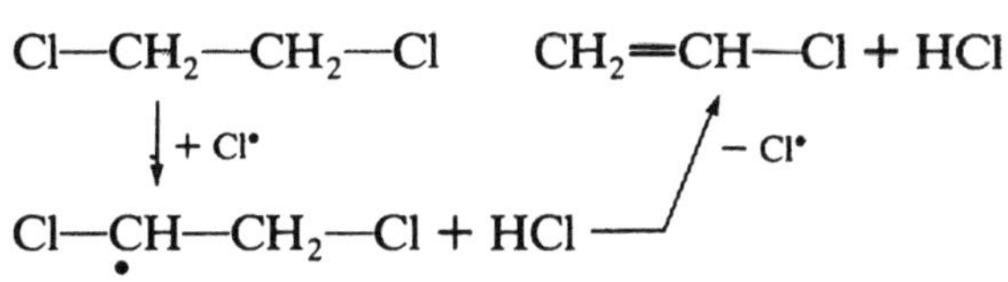

(42)

Since chlorine is rather expensive, all three steps, chlorination, oxychlorination and

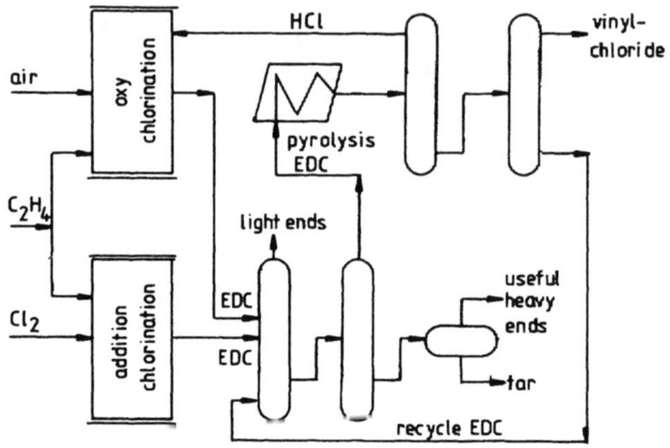

Fig. 8-3. Industrial process for the manufacture of vinyl chloride by combined chlorination and oxychlorination.

dehydrochlorination, are integrated in one process and balanced in such a way that no chlorine is lost. The principle set-up is illustrated in Figure 8.3.

The first industrial processes for the manufacture of vinyl chloride were based on acetylene as the feedstock. HCl was added to acetylene at 140—200 °C in the presence of $HgCl_2$ as the catalyst. The reaction is straightforward and proceeds with high conversions (96%) of acetylene and high selectivities (98%) for vinyl chloride. Nevertheless ethylene has replaced acetylene as the starting material, since it is significantly cheaper.

$$CH{\equiv}CH + HCl \xrightarrow{HgCl_2} CH_2{=}CH{-}Cl \qquad (43)$$

8.7.2. VINYL FLUORIDE

One way to produce vinyl fluoride is by addition of hydrogen fluoride to acetylene. Mercury compounds on a heterogeneous support catalyze the reaction at 40—150 °C, fluorides of alumina, zinc etc. require higher temperatures (250—400 °C).

$$CH{\equiv}CH + HF \xrightarrow[AlF_3,\ ZnF_2]{Hg/supp.} CH_2{=}CHF \qquad (44)$$

In a second method hydrogen fluoride is added to vinyl chloride to give 1-fluoro-1-chlorethane at ambient temperature. In a consecutive step HCl is eliminated with a Cu-catalyst at high temperatures up to 600 °C.

$$CH_2{=}CHCl + HF \rightarrow CH_3{-}CHFCl \qquad (45)$$

$$CH_3{-}CHFCl \xrightarrow[600^\circ C]{[Cu]} CH_2{=}CHF + HCl \qquad (46)$$

Vinyl fluoride is polymerized to polyvinylfluoride, a special polymer.

8.7.3. VINYL ACETATE

Another important vinyl monomer is vinyl acetate, which is produced on a scale of 2 million tons per year world-wide. Its production processes underwent a similar development to that of vinyl chloride. Originally it was made from acetylene by addition of acetic acid at 170—250 °C with zinc acetate on charcoal as the catalyst.

$$HC{\equiv}CH + CH_3COOH \xrightarrow{[cat.]} CH_2{=}CH{-}O{-}\overset{\overset{\displaystyle O}{\|}}{C}{-}CH_3 \qquad (47)$$

Today nearly all of the vinyl acetate is produced from ethylene in an oxyacetylation process. The process resembles the oxychlorination when the overall reaction equation is considered:

$$CH_2{=}CH_2 + CH_3COOH + \tfrac{1}{2}\,O_2 \xrightarrow{[cat.]} CH_2{=}CH{-}O{-}\overset{\overset{\displaystyle O}{\|}}{C}{-}CH_3 + H_2O \qquad (48)$$

ICI and Celanese developed a liquid-phase process, in which the reaction conditions and the catalysts, $PdCl_2$ and CuO, are similar to the Wacker-process for the production of acetaldehyde from ethylene. (See Chapter 5.)

For economic reasons gas-phase processes dominate the vinyl acetate production. Palladium metal or palladium salts are the catalysts. The temperatures are typically 175—200 °C and the pressures 0.5—1 MPa. The gas-phase process is not an addition reaction and is therefore not discussed in detail here.

Vinyl ethers are produced from acetylene even today. Alcohols react with acetylene in the presence of alkali metal hydroxide or alcoholate at 120 to 180 °C and pressures of up to 2 MPa.

$$CH{\equiv}CH + ROH \xrightarrow{[ROK]} CH_2{=}CH{-}OR \qquad (49)$$

R can be methyl, ethyl, propyl, butyl etc. Vinyl ethers find a number of applications as additives to paints, special polymers, adhesives etc.

8.8. Acrylic Acid Derivatives

8.8.1. ACRYLIC ACID

Today most acrylic acid is produced by direct oxidation of propene over heterogeneous heavy metal molybdate catalysts at temperatures between 200 and 500 °C and slightly elevated pressures. The direct oxidation is not acid or base catalyzed and is therefore covered in Chapter 5. However, many of the traditional acrylic acid manufacturing processes involve acid and base catalyzed addition, substitution and elimination processes and should be discussed here in a little more detail.

The first step of the cyanohydrin process is the addition of HCN to ethylene oxide in the presence of a base catalyst. In a second step the hydroxypropionitrile is then treated with alcohol and sulfuric acid in stoichiometric amounts to hydrolyze the nitrile group and to eliminate water simultaneously, yielding the acrylic acid ester as the main product.

$$\underbrace{CH_2{-}CH_2}_{\diagdown O \diagup} + HCN \xrightarrow{[OH^-]} HO{-}CH_2{-}CH_2{-}CN \tag{50}$$

$$ROH + HO{-}CH_2{-}CN_2{-}CN \xrightarrow{[H_2SO_4]} CH_2{=}CH{-}\overset{\overset{O}{\|}}{C}{-}OR + NH_3 \tag{51}$$

The hydrolysis step can, of course, also be applied directly to acrylonitrile which was produced by the SOHIO process (= ammonoxidation of propene).

The Reppe process is an addition of carbon monoxide and water to acetylene in a liquid-phase process with nickel bromide as the catalyst at 160—200 °C and 5.0 MPa. A modern offspring is the reaction of ethylene, CO and oxygen on heterogeneous Pd catalysts. However both processes are not acid or base catalyzed and are mentioned here only to complete the picture.

$$CH{\equiv}CH + CO + H_2O \xrightarrow[\text{homogeneous}]{[Ni]} CH_2{=}CH{-}COOH \tag{52}$$

$$CH_2{=}CH_2 + CO + \tfrac{1}{2}\, O_2 \xrightarrow[\text{heterogeneous}]{[Pd]} CH_2{=}CH{-}COOH \tag{53}$$

Another interesting process is the cycloaddition of formaldehyde to ketene with a Lewis acid catalyst (e.g. $AlCl_3$, BF_3, $ZnCl_2$ etc.) and the consecutive ring opening of the butyrolactone with phosphoric acid and copper as the catalyst.

$$H_2C{=}O + CH_2{=}C{=}O \xrightarrow{[AlCl_3]} \begin{array}{l} CH_2{-}C{=}O \\ \,|\qquad\; | \\ CH_2{-}O \end{array} \tag{54}$$

$$\xrightarrow[{[Cu]}]{+\,H_3PO_4} CH_2{=}CH{-}COOH$$

8.8.2. ACRYLONITRILE

Acrylonitrile is produced industrially by the ammonoxidation of propene, the best-known process being the SOHIO process, which is described in detail in Chapter 5.

$$CH_2{=}CH{-}CH_3 + NH_3 + 1\tfrac{1}{2}\, O_2 \xrightarrow{\text{cat.}} CH_2{=}CH{-}CN + 3\, H_2O \tag{55}$$

Before this modern oxidation process was introduced in the 1950s, a variety of acid or base catalyzed addition processes were used. The base catalyzed addition of HCN to ethylene oxide was already mentioned in connection with the manufacture of acrylic acid.

$$\underset{\text{O}}{CH_2\text{—}CH_2} + HCN \xrightarrow[\text{[OH}^-\text{]}]{} HO\text{—}CH_2\text{—}CH_2\text{—}CN \quad (56)$$

$$HO\text{—}CH_2\text{—}CH_2\text{—}CN \xrightarrow{Al_2O_3} CH_2\text{=}CH\text{—}CN + H_2O \quad (57)$$

Acetylene is another possible feedstock for acrylonitrile, since HCN can be added in a $CuCl/NH_4Cl$ catalyzed reaction, which proceeds at 80—90 °C:

$$HC\equiv CH + HCN \xrightarrow[NH_4Cl]{CuCl} CH_2\text{=}CH\text{—}CN \quad (58)$$

Another possibility is the addition of HCN to acetaldehyde leading to lactic nitrile, which then is dehydrated with phosphoric acid to give acrylonitrile.

$$CH_3\text{—}C(\text{=O})H + HCN \rightarrow CH_3\text{—}C(OH)(H)\text{—}CN \xrightarrow[-H_2O]{} CH_2\text{=}CH\text{—}CN \quad (59)$$

2.5 Million tons of acrylonitrile are produced annually. The major uses are as acrylic fibers (65%) and as comonomer for the production of acrylonitrile-butadiene-styrene and styrene-acrylonitrile copolymers (15%).

8.8.3. ALLYL COMPOUNDS

Allyl chloride is the starting material for several addition and substitution processes of industrial importance. It is manufactured via a free radical chlorination of propene or by the oxychlorination of propene with $PdCl_2$—$CuCl_2$ catalyst.

$$CH_2\text{=}CH\text{—}CH_3 + Cl_2 \xrightarrow{500°C} CH_2\text{=}CH\text{—}CH_2Cl + HCl \quad (60)$$

$$CH_2\text{=}CH\text{—}CH_3 + HCl + \tfrac{1}{2}\,O_2 \xrightarrow[\text{[CuCl}_2\text{]}]{\text{[PdCl}_2\text{]}} CH_2\text{=}CH\text{—}CH_2Cl + H_2O \quad (61)$$

The addition of HOCl to allyl chloride leads to a mixture of the dichlorohydroxypropanes, which are dehydrochlorinated with base to give *epichlorohydrin.*

$$CH_2\text{=}CH\text{—}CH_2Cl + HOCl \xrightarrow{[H^+]} \underset{Cl\quad OH\quad Cl}{CH_2\text{—}CH\text{—}CH_2}\ 70\% + \underset{OH\quad Cl\quad Cl}{CH_2\text{—}CH\text{—}CH_2}\ 30\% \quad (62)$$

$$\xrightarrow[\substack{-\,CaCl_2\\ -\,2\,H_2O}]{+\,Ca(OH_2)} \underset{\text{O}}{CH_2\text{—}CH}\text{—}CH_2\text{—}Cl$$

About 50% of the 0.6 million tons of epichlorohydrin produced annually is used for the production of epoxy resins. The first step is the base catalyzed addition of bisphenol A to epichlorohydrin; the intermediate product rapidly looses HCl to from a bisglycidyl ether. The ether may react with further bisphenol A. The molecular weight of the resulting macromolecule can be adjusted by selecting the molar ratio of epichlorohydrin to bisphenol A. Base catalyzed addition of polyalcohols or polyamines leads to epoxy resins which find many applications as glues, binders, for surface protection, as casting resins, etc.

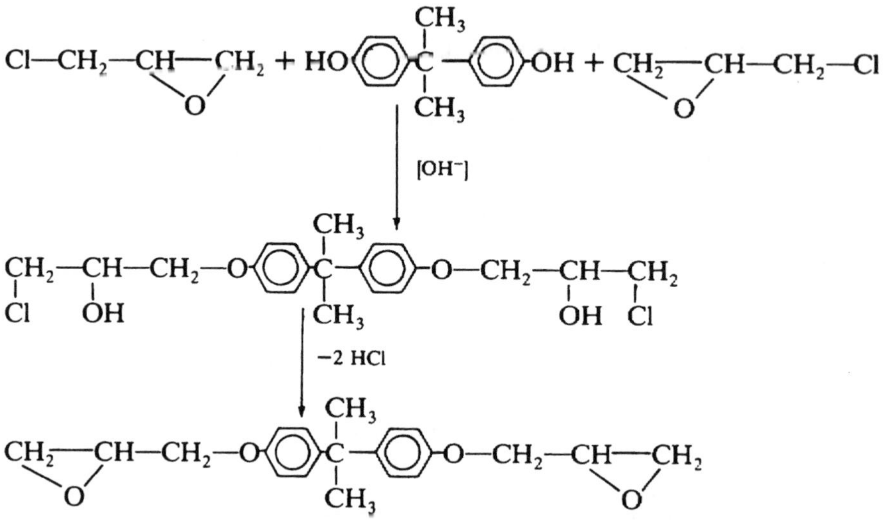

(63)

Glycerol is another important chemical, which is produced from allyl compounds by addition and substitution processes. The oldest and still most important route to synthetic glycerol starts with epichlorohydrin. It is hydrolyzed in the first step with diluted sodium hydroxide solution. The second step is the dehydrochlorination, followed by another hydrolysis of the epoxy intermediate.

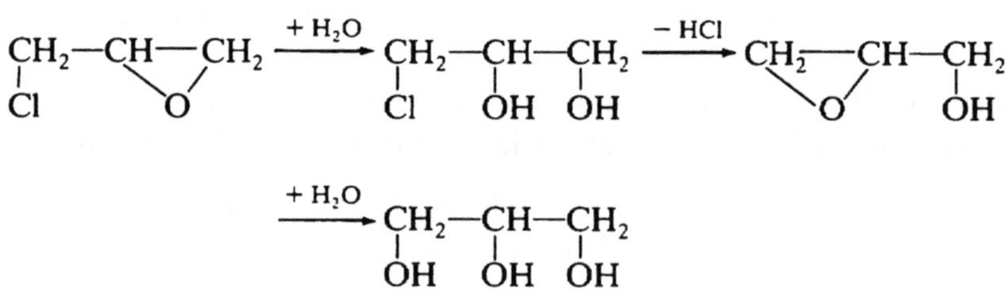

(64)

The valuable chlorine can be saved when allyl alcohol reacts with hydrogen peroxide or peracetic acid and the resulting epoxide is hydrolyzed to give glycerol

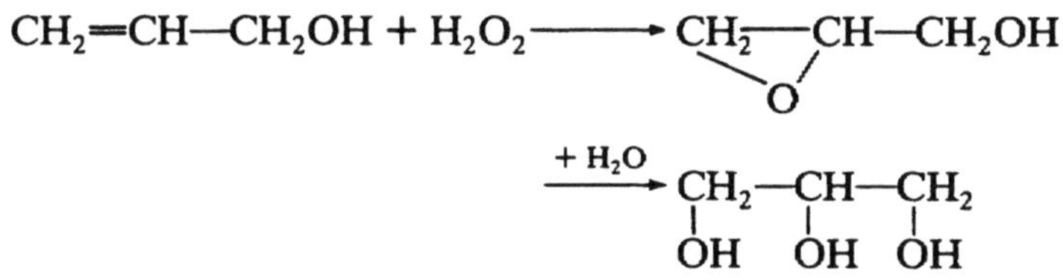
(65)

However, even today, only about 30% of the 0.5 million tons of glycerol produced annually is synthetic. Seventy percent is produced by saponification of natural fatty acid triglycerides. The major uses of glycerol are for the production of polyurethanes, alkyd resins and in the food, cosmetics, and detergent industries.

8.8.4. METHACRYLIC ACID DERIVATIVES

Methyl methacrylate is used for the production of glass-like plastics, known under the commercial name of Plexiglas®. It can be produced from acetone by a base-catalyzed hydrocyanation. The resulting cyanohydrin is then converted to methyl methacrylate with methanol and sulfuric acid.

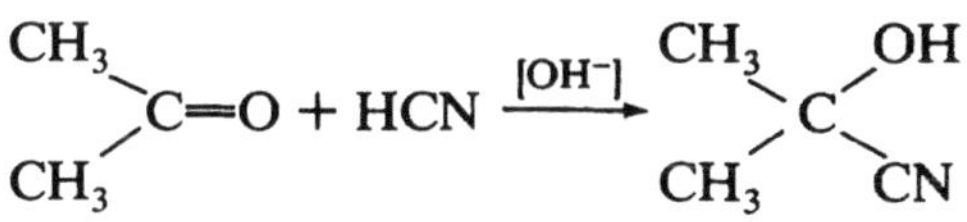
(66)

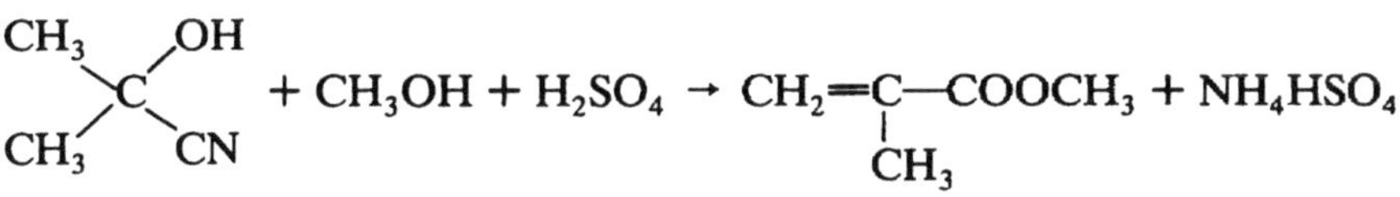
(67)

Recent work has aimed at the development of alternatives which avoid the formation of the ammonium hydrogen sulfate, but a real breakthrough has not been achieved so far.

8.8.5. MISCELLANEOUS REACTIONS

Methyl isobutyl ketone (MIBK) is the major consecutive product of acetone. The first step is a base-catalyzed aldol condensation between two molecules of acetone. The second step is an acid catalyzed elimination of water to give mesityl oxide which is also an important intermediate product. Its partial hydrogenation then yields MIBK.

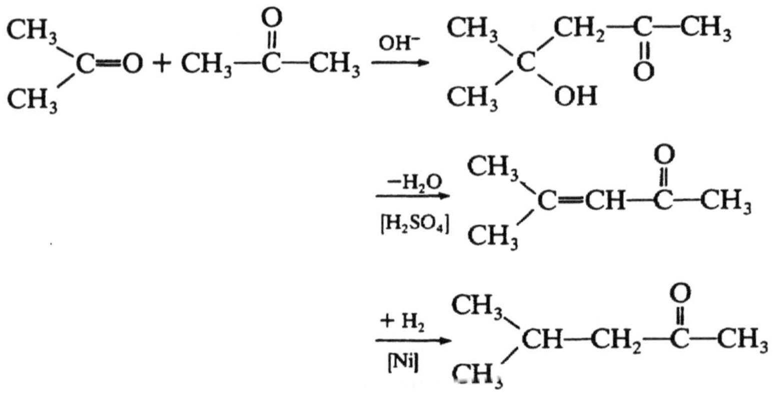

(68)

Multifunctional heterogeneous acidic catalysts, doped with platinum or palladium have been tested successfully for the formation of MIBK from acetone in a one-step process, in which all the above mentioned reactions occur in one catalyst bed. MIBK is an important solvent for instance in the paint industry.

Adiponitrile is a very important starting material in the polyamide industry. Numerous processes are known for its production. Only those which are acid or base catalyzed are mentioned here.

Starting from adipic acid and ammonia the initially formed diamonium salt is dehydrated in two steps in acid catalyzed reactions to adiponitrile.

$$\mathrm{HOOC{-}(CH_2)_4{-}COOH} \xrightarrow{+\,2\,\mathrm{NH_3}} \mathrm{H_4NOOC{-}(CH_2)_4{-}COONH_4}$$

$$\xrightarrow[\mathrm{H_3PO_4}]{-\,2\,\mathrm{H_2O}} \mathrm{N_2NOC{-}(CH_2)_4{-}CONH_2}$$

$$\xrightarrow[\mathrm{H_3PO_4}]{-\,2\,\mathrm{H_2O}} \mathrm{NC{-}(CH_2)_4{-}CN} \qquad (69)$$

Butadiene can also serve as a precursor for adiponitrile. A free radical gas-phase chlorination of butadiene at 200—300 °C yields all possible dichlorobutenes. The isomeric mixture is then treated with HCN or NaCN in the presence of copper cyanide as a homogeneous catalyst. Due to an allyl rearrangement during the cyanation the 1,4-dicyanobutene is formed in 95% selectivity.

$$\mathrm{CH_2{=}CH{-}CH{=}CH_2} \xrightarrow{+\,\mathrm{Cl_2}} \mathrm{CH_2{=}CH{-}CHCl{-}CH_2Cl} +$$

$$+\ \mathrm{ClCH_2{-}CH{=}CH{-}CH_2Cl}$$

$$\xrightarrow[-\,\mathrm{Cl^-}\ [\mathrm{cat.}]]{+\,\mathrm{CN^-}} \mathrm{NC{-}CH_2{-}CH{=}CH{-}CH_2{-}CN}$$

$$\xrightarrow[[\mathrm{Pd}]]{\mathrm{H_2}} \mathrm{NC{-}(CH_2)_4{-}CN} \qquad (70)$$

DuPont developed a one-step hydrocyanation process of butadiene. Nickel(0) phosphines with aluminum and zinc promoters constitute the catalyst system in a homogeneous liquid phase at normal pressure and elevated temperature.

$$CH_2{=}CH{-}CH{=}CH_2 + 2\ HCN \xrightarrow[\text{[catal.]}]{} NC{-}(CH_2)_4{-}CN \tag{71}$$

References

Tanabe, K.: *Solid Acid and Bases*, Kodanska Tokyo, Academic Press, New York, 1970.

Tanabe, K.: 'Solid Acid and Base Catalysts' in *Catalysis Science and Technology*, Anderson, J. R. and Boudart, M. editors, Vol. 2, pp. 231—73, Springer, Berlin, 1981.

Tanabe, K. (guest editor): *Advances in Acidic and Basic Solid Materials*, Elsevier Sequoia, 1987.

Imelik, B. (editor): *Catalysis by Acids and Bases*, Elsevier Sequoia, Amsterdam, 1985.

Finston, H. L. and Rychtman, A. C.: *A New View of Current Acid-Base Theories*, Wiley, New York, 1982.

Beranek, L. and Kraus, M.: 'Heterogeneous Eleminations, Additions, and Substitutions' in *Comprehensive Chemical Kinetics*, Bamford, C. H. and Tipper, C. F. H. editors, Vol. 20, p. 263 ff. Elsevier, Amsterdam, 1978.

Weissermel, K. and Arpe, H. J.: *Industrial Organic Chemistry*, Verlag Chemie, Weinheim—New York, 1978.

CHAPTER 9

Catalytic Polymerization

9. Introduction

Organic polymers are long-chain compounds made of several repeating units. These polymers are either natural, such as protein, cellulose, starch or synthetic, such as polyethylene (a plastic), polyesters and polyamides (plastics and fibers) and polybutadiene (rubber).

Many organic compounds possessing multiple bonds have the tendency to polymerize in the presence of an initiator. Polymerization of these compounds can either be affected by the use of chemical initiators such as free radicals and some ionic compounds or by irradiation. Visible, ultraviolet, X-ray, γ-ray irradiation etc. could be used to start free radical polymerizations.

Ionic-type initiators are customarily termed catalysts, although in the true sense they are not, since the initiator species are not fully recovered. However, the words 'catalyst' and 'initiator' will be used interchangeably in the course of our discussion.

Condensation polymerization of compounds having at least two functional groups is usually initiated with true catalysts such as Brønsted and Lewis acids. An example of this polymerization type is the condensation of a dicarboxylic acid and a diol which produces a polyester.

$$\text{HOROH} + \text{HOOCRCOOH} \rightleftharpoons -\!\!\left[\text{OROC}(=\text{O})\text{RCO}\right]\!\!-_n + H_2O$$

The building units for polymers are simple molecules which are termed monomers. Polymers represent the end products of a great number of monomers produced via different petrochemical routes.

Commercial polymers add to the comfort of human beings. New industries have been built based on polymers. Many products of a natural origin (cotton, wool, iron, aluminum etc.) have been either partially or totally replaced by different polymeric materials, produced from petrochemical sources.

9.1. Monomers, Polymers and Copolymers

A monomer is a reactive compound, having one or more functional groups. It can easily be initiated or catalyzed to react with itself or with a different monomer, forming a long chain of repeating units. When the same monomer is used, the polymer is called a homopolymer. For example, when ethylene is polymerized, polyethylene is formed from repeating ethylene units.

$$n\,CH_2{=}CH_2 \rightarrow -\!\!\left[CH_2{-}CH_2\right]\!\!-_n$$

If ethylene is polymerized, however, in presence of another monomer such as propylene, a copolymer is formed which may have the structure shown:—

$$n\,CH_2{=}CH_2 + n\,CH_2{=}CHCH_3 \rightarrow \left[CH_2CH_2CH_2\underset{}{\overset{CH_3}{\overset{|}{C}}}H\right]_n$$

A copolymer is a polymer made from two or more different monomers. A block copolymer is produced by adding another monomer to an active polymeric species. For example, a block of styrene could be attached to a block of isoprene.

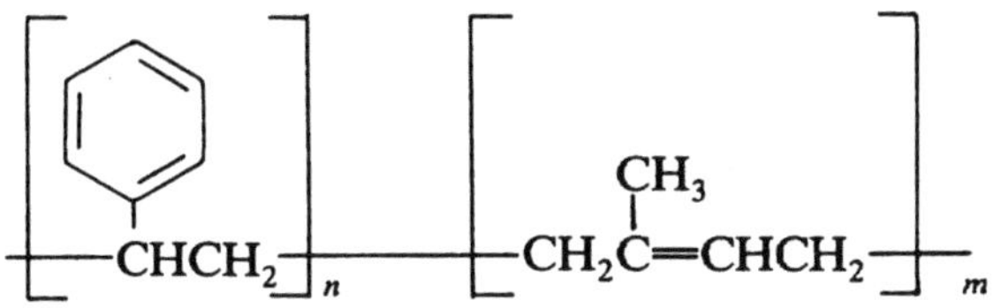

Alternating copolymers have the monomers of one type alternating in a regular manner with the monomers of the other type, as in the case of styrene-butadiene.

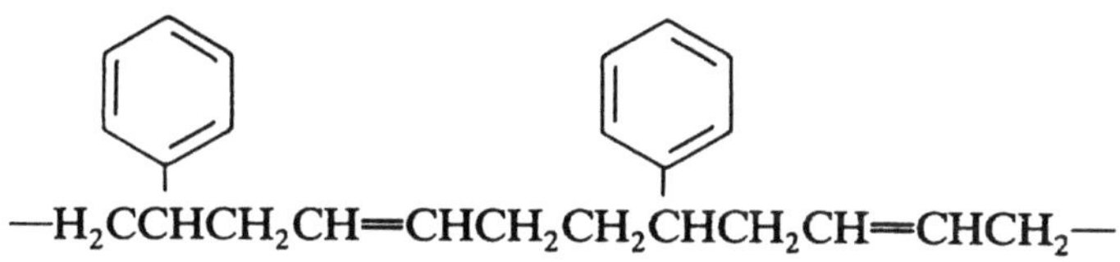

Random copolymers have different monomer molecules distributed in a random manner along the polymer chain.

Like polymers, some copolymers are of great commercial importance, such as the copolymer of butadiene and styrene, which is the most widely used synthetic rubber. These types of polymers are sometimes called addition polymers, since they react by an addition-type reaction.

Another type of polymer is produced by a condensation reaction between the same monomers or between different monomers. These are termed condensation polymers or step polymers since the reaction takes place in steps. An example of a condensation polymer produced from the same monomer is nylon 6, which may be synthesized from caprolactam. Caprolactam is a 7-membered ring compound which can open the ring under certain conditions to give an amino acid. Because it is bifunctional, self condensation takes place to give nylon 6.

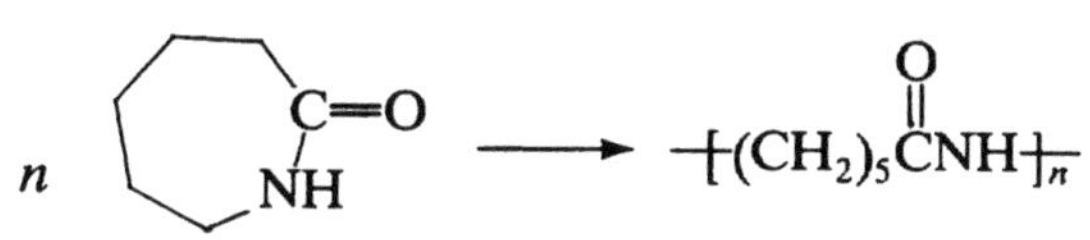

If, on the other hand, a monomer having at least two similar functional groups that can react with the functional groups of a different monomer having also two similar functional groups, a condensation type polymer is formed. For example terephthalic acid has two carboxyl groups that can condense with the hydroxyl groups of another monomer such as ethylene glycol, and the product is polyethylene terephthalate.

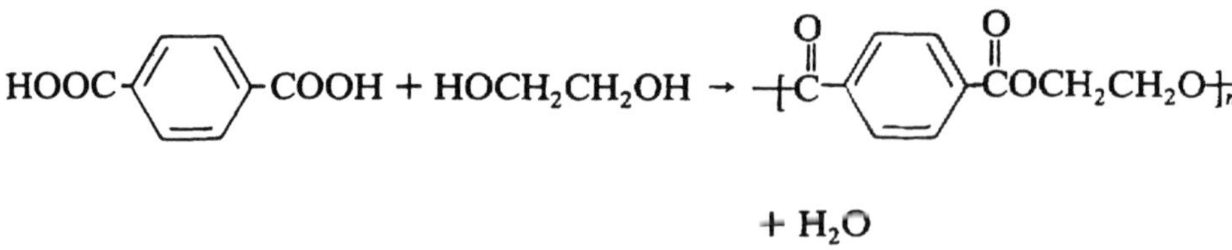

In certain cases, the polymerization (by addition) may be limited to a dimer when two similar molecules react together and stop at this stage, or to a trimer, which is formed by reacting the dimer with a similar monomer, or to a tetramer etc. These polymeric materials have a much lower average molecular weight than high molecular weight polymers and are termed oligomers. Oligomers are not always linear molecules but are sometimes cyclic compounds. For example, trioxane is a trimer of formaldehyde, and it is a six-membered ring compound. Formaldehyde can also form polyformaldehyde, which is a linear oligomer.

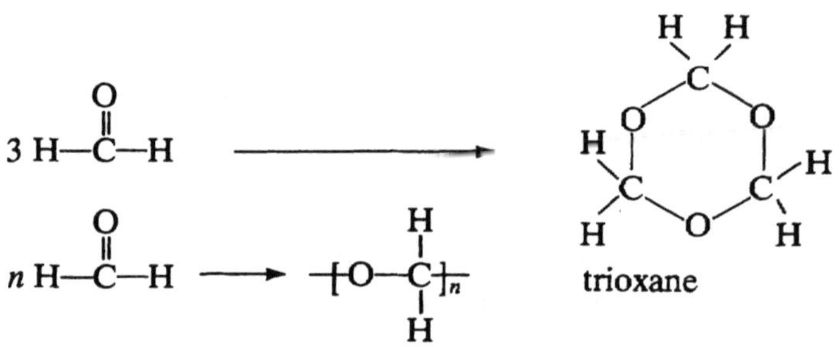

trioxane

polyformaldehyde

Polymers could also be described as being linear or branched. A linear polymer consists of repeating monomer units with no branches originating from the polymer chain. Linear polyethylene and polypropylene are examples of this type. The methyl group in polypropylene however, is not considered a branch. If a side chain originates from the polymer backbone the polymer is branched.

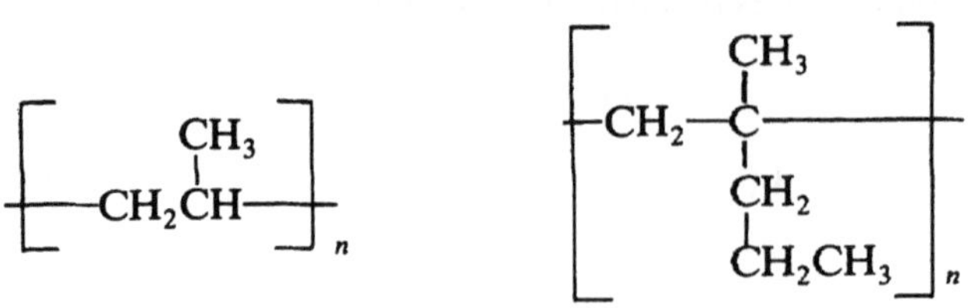

Linear polypropylene Branched polypropylene

Crosslinked polymers are those in which two or more polymer chains are linked together at one or more points other than at their ends. This may form a network structure and these polymers are sometimes called network polymers. If cross-linking is high, the polymer may be rigid with a high melting point. Light cross-linking produces a polymer which has elastomeric properties. Crosslinking may be formed during the polymerization when monomers with multifunctional groups are polymerized as with phenol-formaldehyde resins or by the use of outside linking agents as in the vulcanization of rubber.

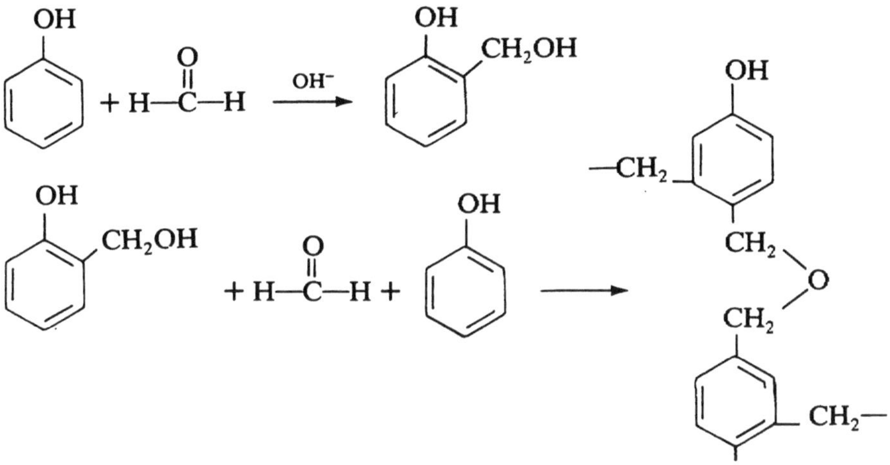

Crosslinked phenol-formaldehyde

Polymers are generally nonhomogeneous and will not show a definite temperature at which the polymer crystallizes. Two temperatures are used to describe the effect of heating on a polymeric material.

The crystalline melting point (T_m), is the temperature at which a polymer's crystallites disappear or melt. Crystallites will appear when a melted polymer is cooled to T_m. If further cooling is carried out, the amorphous regions change to a glass-like material and this temperature is the glass transition temperature (T_g).

9.2. Classification of Polymers

Synthetic polymers are classified in several ways. The following classifications are used:—

Chemical type of monomer: Polyolefins, polyesters, polyamides etc.

Type of polymerization reaction: Addition and condensation polymers.

Type of polymerization technique used: Bulk, solution, suspension, emulsion.

Type of utilization: Thermosetting, thermoplastic, fibers, elastomers, engineering plastic etc.

Polymers are generally distinguished by their end uses into plastics, synthetic fibers and elastomers.

9.2.1. PLASTICS

Plastics are relatively tough substances of high molecular weight. Generally plastics are subclassified into thermoplastics and thermosets. Thermoplastics soften when heated while thermosetting plastics are highly crosslinked solids which do not soften upon heating.

Thermoplastics may be amorphous, partially crystalline or crystalline. Table 9-1 shows the structure of the major thermoplastic polymers.

9.2.2. SYNTHETIC FIBERS

Synthetic fibers are polymers formed mainly by condensation polymerization. However some synthetic fibers are also synthesized by addition polymerization. Fibers are much less elastic than plastics and elastomers. They undergo low elongation and are characterized by their light weight, high tensile strength, high resistance to deformation and low moisture absorption. Polymers for fibers are highly crystalline due to secondary forces. The most important fiber-forming polymers are polyesters and polyamides which are made by condensation reactions, and polyacrylics and polypropylene made by addition polymerization.

9.2.3. ELASTOMERS

Elastomers differ from plastics and fibers in the presence of residual double bonds in the backbone of the polymer. Unsaturation provides some centers for weak crosslinking between the linear chains which prevents them from slipping over one another. Polymers of this type are predominantly amorphous, having a high degree of reversible elasticity. The most important monomers for elastomer production are butadiene, isoprene, chloroprene (conjugated diolefins), and isobutylene, cyclopentene and propylene (monoolefins).

9.3. Polymerization Reactions

Polymerization can take place either by condensation or by addition. The first proceeds by a stepwise mechanism and polymers formed are sometimes called step polymers. Addition polymerization proceeds by a chain-type reaction. This is the most widely used polymerization scheme for producing most synthetic rubbers, plastics and some synthetic fibers. A special type of polymerization takes place by a ring opening reaction. Some of the most important commercial polymers are prepared by this method. Since either an addition or a condensation reaction precedes the ring opening, this polymerization will be discussed in a separate section.

Table 9-1. Major thermoplastic polymers.

Name	Abbreviation	Family	Formula	Melting temp.	Density
Low density polyethylene	LDPE	Polyolefin	CH_2-	110 (T_m)	0.910
High density polyethylene	HDPE	Polyolefin	CH_2-	120 (T_m)	0.950
Polypropylene	PP	Polyolefin	$-CH_2\text{-}CH(CH_3)-$	175 (T_m)	0.902
Polyvinylchloride	PVC	Vinyl	$-CH_2\text{-}CH(Cl)-$	100 (T_g)	1.35
Polyvinyl acetate	PVA	Vinyl	$-CH_2\text{-}CH(O\text{-}C(=O)CH_3)-$		
Polystyrene	PS	Styrenic	$-CH_2\text{-}CH(C_6H_5)-$	100 (T_g)	1.05
Acrylonitrilebutadiene styrene	ABS	Styrenic			
Acrylonitrile styrene	SAN	Styrenic			
Polymethylmethacrylate		Acrylic	$-CH_2\text{-}C(C(=O)OCH_3)(CH_3)-$		
Polyhexamethylenediamide	Nylon 66	Polyamide	$-NH\text{-}(CH_2)_6\text{-}NH\text{-}C(=O)\text{-}(CH_2)_4\text{-}C(=O)-$	265 (T_m)	1.14

Table 9-1. (continued)

Name	Abbreviation	Family	Formula	Melting temp.	Density
Polycaprolactam	Nylon 6	Polyamide	$-NH-(CH_2)_5-\overset{O}{\overset{\parallel}{C}}-$	225 (T_m)	1.14
Polyethyleneterephthalate	PET	Polyester	$-\overset{O}{\overset{\parallel}{C}}-C_6H_4-\overset{O}{\overset{\parallel}{C}}-OCH_2CH_2O-$	270 (T_m)	
Polybutyleneterephthalate	PBT	Polyester	$-\overset{O}{\overset{\parallel}{C}}-C_6H_4-\overset{O}{\overset{\parallel}{C}}-O-(CH_2)_3-CH_2-O-$	250 (T_m)	1.3
Polycarbonates	PC	Polyester	$-O-R'-O-\overset{O}{\overset{\parallel}{C}}-O-R-O-\overset{O}{\overset{\parallel}{C}}-$	190 (T_g)	1.2
Polyacetais		Polyethers	$-O-CH_2-O-R-$	181 (T_m)	

9.3.1. CONDENSATION POLYMERIZATION

In general, condensation or step polymerization takes place by a reaction between two monomers with the elimination of a small molecule such as water. In some cases the polymerization can take place between one monomeric species having two functional groups as in the case of polymerizing a hydroxycarboxylic acid or an amino acid. In polymerizations of this type, the monomer disappears much faster than in addition polymerizations. A dimer is first formed which may react with another monomer forming a trimer. A tetramer is formed when a trimer reacts with a monomer, or when a dimer reacts with another dimer. A pentamer (and higher oligomers) is formed in the same way in a stepwise sequence. Although step polymerizations are generally slower than addition polymerizations and long reaction times are needed for high conversions, the monomers disappear fast. During the course of the reaction, dimers, trimers, tetramers etc. are present in the reaction medium and a high molecular-weight polymer is not obtained until the end of the polymerization. The most important polymers of this type are polyesters, polyamides, the various condensation polymers of formaldehyde, polycarbonates, polyurethanes and polyethers.

Condensation reactions are generally catalyzed with acid type catalysts such as metal oxides and sulfonic acids. These catalysts speed up the reaction by coordination with the carbonyl group. In polyesterification, for example, the nucleophilic species (the diol) adds to the carbonyl carbon followed by elimination of water.

$$\mathrm{HOOCRCOOH + HOR'OH \rightarrow -\!\!\left[\overset{\displaystyle O}{\overset{\|}{C}}R\overset{\displaystyle O}{\overset{\|}{C}}OR'O\right]\!\!-_{n} + H_2O}$$

In forming polyamides, the reaction may take place between a diacid and a diamine.

$$\mathrm{HOOCRCOOH + H_2NR'H_2 \rightarrow -\!\!\left[\overset{\displaystyle O}{\overset{\|}{C}}R\overset{\displaystyle O}{\overset{\|}{C}}NHR'NH\right]\!\!- + H_2O}$$

Polyamides may also be formed by polymerization of an amino acid or by the ring opening of a lactam.

The carboxyl group of one molecule reacts with the amino group of another molecule and so on until a polymer is formed.

$$n\,\mathrm{H_2NRCOOH \rightarrow -\!\!\left[R\overset{\displaystyle O}{\overset{\|}{C}}NH\right]\!\!-_{n} + H_2O}$$

9.3.2. ADDITION POLYMERIZATION

Addition polymerization is mainly used for polymerizing olefins, conjugated diolefins, and some carbonyl compounds. Free radical, anionic or cationic initiators may be used to induce the reactions. Coordination compounds such as Ziegler-type catalysts are important initiators which are sometimes classified with the anionic types.

The general scheme by which the initiation and propagation reactions take place during the polymerization using these chemical initiators bear some similarities.

The first step would be the addition of a free radical, an anion or a cation initiator (I^z) to an olefin. When an unsymmetrical olefin ($CH_2{=}CHR$) is used, the addition would predominantly be to the olefin side that produces the most stable intermediate (A).

$$I^z + CH_2{=}CHR \rightarrow [I{-}CH_2{-}\underset{\displaystyle R}{\underset{|}{C}}H]^z \quad \text{(A)}$$

$$I^z + CH_2{=}CHR \rightarrow [I{-}\underset{\displaystyle R}{\underset{|}{C}}HCH_2]^z \quad \text{(B)}$$

I^z = a free radical $I^\bullet$, a cation I^+, or an anion I^-. R = alkyl, aryl, Cl etc.

Propagation then takes place by a successive addition of the monomer molecules to the intermediate. Three addition modes are possible.

Head-to-tail mode which gives the most stable intermediate (A′)

$$[I{-}CH_2{-}\underset{\displaystyle R}{\underset{|}{C}}H]^z + CH_2{=}CHR \rightarrow [I{-}CH_2{-}\underset{\displaystyle R}{\underset{|}{C}}H{-}CH_2{-}\underset{\displaystyle R}{\underset{|}{C}}H]^z \quad \text{(A')}$$

Head-to-head and tail-to-tail modes are less likely to occur since the intermediate (C) from the former case is less stable than (A′) and in the latter case (D) is produced from the less stable intermediate (B).

$$[I{-}CH_2{-}\underset{\displaystyle R}{\underset{|}{C}}H]^z + CH_2{=}CHR \rightarrow [I{-}CH_2{-}\underset{\displaystyle R}{\underset{|}{C}}H{-}\underset{\displaystyle R}{\underset{|}{C}}HCH_2]^z \quad \text{(C)}$$

$$[{-}\underset{\displaystyle R}{\underset{|}{C}}HCH_2]^z + CH_2{=}CHR \rightarrow [I{-}\underset{\displaystyle R}{\underset{|}{C}}HCH_2CH_2CHR]^z \quad \text{(D)}$$

Chain growth continues until termination takes place. An important difference between free radical and ionic polymerization types is the absence of a counter ion in the former case. Presence of a counter ion in ionic-type addition polymerizations has a great effect on the reaction sequence, the concentration of the reactants and on the termination steps.

If, on the other hand, a symmetrical olefin is polymerized, such as ethylene or 2-butene, only one mode of addition is possible. For example polymerizing ethylene should produce linear polyethylene. However another phenomenon takes place which is responsible for branching due to a side reaction, or to the chain transfer reaction discussed below.

9.3.2.1. *Free Radical Addition Polymerization*

Many polymerization reactions can be started by the use of free radicals. Free radical initiators are compounds possessing a weak covalent bond that can easily break into two free radicals by the action of heat. Peroxides, hydroperoxides, or azo compounds are examples of this type.

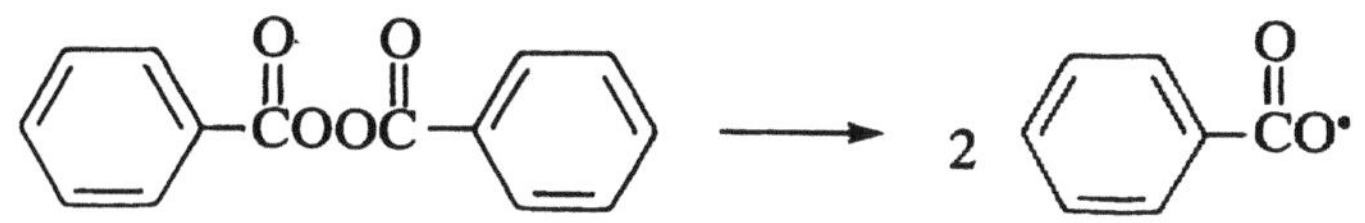

Benzoyl peroxide

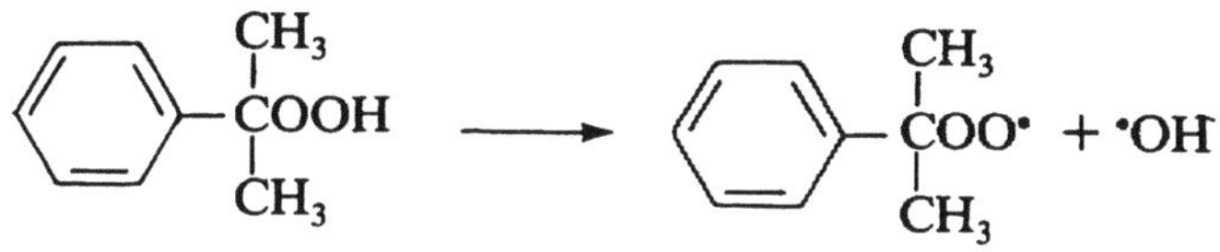

Cumene hydroperoxide

$$(CH_3)\overset{CN}{\overset{|}{C}}H{=}N\overset{CN}{\overset{|}{C}}(CH_3)_2 \rightarrow 2\ (CH_3)_2\overset{CN}{\overset{|}{C}}{}^{\bullet} + N_2$$

Free radical reactions are generally less selective than ionic polymerizations. The polymers produced are branched and therefore have a low crystallinity. For example, low density polyethylene (LDPE), produced by free radical initiators, has a lower T_m (approx. 105 °C) than high density polyethylene (HDPE) produced by ionic or coordination type catalysts (T_m approx. 125 °C). Branching can take place by a chain transfer reaction when a propagating species abstracts a hydrogen atom from another polymer molecule. For example, branching in polyethylene can be represented as:

$$—CH_2—CH_2— + —CH_2\dot{C}H_2 \rightarrow CH_2\dot{C}H— + —CH_2CH_3$$

$$—CH_2\dot{C}H— + CH_2{=}CH_2 \rightarrow —CH_2\underset{\underset{\underset{\bullet}{CH_2}}{|}}{\underset{CH_2}{\underset{|}{C}}}H—$$

A branched propagating polyethylene

Free radical polymers may be terminated by combining with another free radical such as another propagating chain. In this case the tail-to-tail combination is the most likely addition mode.

Termination may also take place by a disproportionation reaction, where a propagating polymer abstracts an atom from another propagating free radical. In addition to the terminated polymer, a terminal olefinic species will appear which may be induced by an initiator to start the reaction again.

$$-\underset{R}{\overset{\bullet}{C}}- + \overset{\bullet}{C}H_2\underset{R}{C}H- \rightarrow -\underset{|}{C}HR + CH_2{=}\underset{R}{C}-$$

Certain compounds can act as chain transfer agents, which may be intentionally added to the reaction medium. For example, carbon tetrachloride can be used to terminate a propagating polymer chain, and become changed to a free radical that can start a new polymerization reaction.

$$-CH_2\underset{R}{\overset{\bullet}{C}}H + CCl_4 \rightarrow -CH_2\underset{R}{C}HCl + \overset{\bullet}{C}Cl_3$$

$$\overset{\bullet}{C}Cl_3 + CH_2{=}CHR \rightarrow Cl_3CCH_2\overset{\bullet}{C}HR$$

9.3.2.2. *Cationic Polymerization*

Cationic polymerizations can be started by the use of strong protonic or Lewis acids. When Lewis acids are the catalysts, a small amount of a proton donating compound such as water or methyl alcohol is added. A complex is formed which provides the proton for the addition reaction. For example, when $AlCl_3$ is used with water the complex may be represented as:

$$Cl_3AlOH_2 \rightleftharpoons [Cl_3AlOH]^- + H^+$$

Cationic polymerizations are favored when monomers possess an electron-donating group that can stabilize the carbonium ion formed, or the ion-pair intermediate. For example, isobutylene and 1,3-butadiene are typical monomers of this type. Aldehydes could also be polymerized with cationic initiators. Cationic polymerizations are not used as much as free radical or anionic polymerizations. It is a technique that is used more often when control is not necessary, as in the case of phenol-formaldehyde resins and isobutylene polymers. The polymerization of dienes with cationic initiators gives low molecular-weight polymers and cyclic compounds are formed. Polymerization is usually performed in bulk (without the use of a solvent) or in solution. Solvents used must be unreactive, otherwise chain transfer may be dominant.

Low temperature conditions are needed to avoid side reactions. As was mentioned before, addition should give the most stable carbonium ion intermediate. In contrast to free radical propagation (where no counter ion is present), propagation in cationic polymerization involves the insertion of the monomer molecules between the ion-pair. For example, when a proton acid is used with isobutene, initiation and propagation steps could be represented as:—

$$HX + H_2C{=}\overset{CH_3}{\overset{|}{C}}-CH_3 \rightarrow (CH_3)_3C^+X^- \qquad \text{Initiation}$$

$$(CH_3)_3C^+X^- + CH_2{=}\overset{\overset{\displaystyle CH_3}{|}}{C}{-}CH_3 \rightarrow (CH_3)_3CCH_2\overset{+}{C}(CH_3)_2X^- \qquad \text{Propagation}$$

$$(CH_3)_3CCH_2\underset{\underset{\displaystyle CH_3}{|}}{\overset{\overset{\displaystyle CH_3}{|}}{C}}{}^+X^- + n\, CH_2{=}\overset{\overset{\displaystyle CH_3}{|}}{C}{-}CH_3$$

$$\rightarrow (CH_3)_3C{-}[CH_2\overset{+}{C}(CH_3)_2]_{n+1}X^-$$

In cationic polymerization, chain transfer reactions do not take place to an appreciable extent if low temperature-conditions are used. Chain transfer reactions are responsible for decreasing the chain length of the polymer by disproportionation

$$-CH_2\overset{+}{C}(CH_3)_2X^- + CH_2{=}\overset{\overset{\displaystyle CH_3}{|}}{C}{-}CH_3 \rightarrow -CH_2{-}\overset{\overset{\displaystyle CH_3}{|}}{C}{=}CH_2 + (CH_3)_3C^+X^-$$

Termination can take place by the reaction of the propagating polymer with a hydroxy compound such as water.

$$-CH_2\overset{+}{C}(CH_3)_2X^- + H_2O \rightarrow -CH_2\overset{\overset{\displaystyle OH}{|}}{C}(CH_3)_2 + HX$$

9.3.2.3. *Anionic Polymerization*

Many anionic initiators have been used to start the polymerization reactions. Among these are alkyl and aryl lithium, sodium and lithium suspensions in liquid ammonia, aluminum alkyls, sodium naphthalenide etc. Anionic polymerization can be used to polymerize styrene, methylmethacrylates, butadiene etc. The presence of electron withdrawing substituents in the monomer improves the polymerization by stabilizing the intermediate. As in cationic polymerization, a counter ion is always present with the propagating chain. Using *n*-butyllithium as the initiator, and styrene as the monomer, the polymerization sequence could be represented as:—

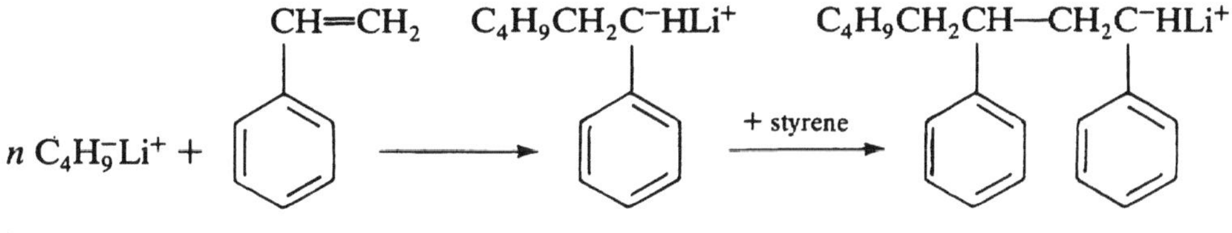

The anion attacks the carbon having the lowest electron density. The intermediate is stabilized by resonance due to the presence of the phenyl group. Propagation takes place by a nucleophilic addition of the anion to the monomer, which becomes inserted between the ion pair.

Anionic polymerization is normally conducted at low temperatures in order to avoid branching due to chain transfer reactions. The reaction can be terminated by the addition of reagents such as water, carbon dioxide, alcohols etc.

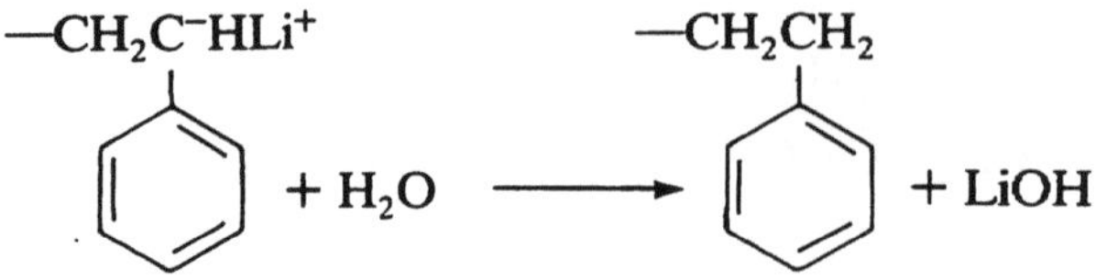

Conjugated diolefins may be polymerized by the use anionic initiators. For example, polymerization of butadiene or isoprene with *n*-butyllithium produces a high *cis*-1,4-polymer. The polymer obtained from isoprene is almost quantitatively the *cis*-1,4-polyisoprene. This has been attributed to the ability of the lithium alkyls to hold isoprene molecules in the cisoid conformation by a π-complex formation

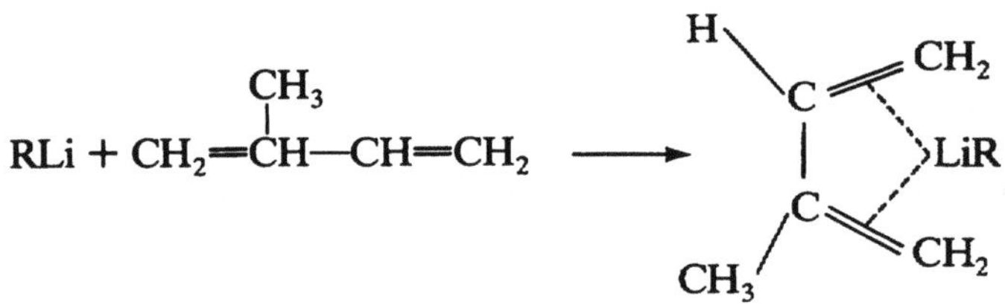

9.3.2.4. *Polymerization with Coordination Compounds*

Ziegler-Natta Catalysts. In ionic polymerization a counter ion is always present to balance the charged propagating polymer. These ions are usually not closely associated with one another. In coordination catalysts the bonds are appreciably covalent. However a certain percentage of ionic character is present and such polymerizations are sometimes classified with the anionic types. Among the catalysts used are the oxides of vanadium, titanium, cobalt and nickel, supported on alumina or charcoal. These catalysts can form π complexes with ligands such as ethylene and butadiene. Another type of coordination catalysts is the π-allyl complexes such as $(\pi\text{-}C_4H_7)_2Ni$ and $(\pi\text{-}C_4H_7)_4Ti$.

These complexes can initiate the polymerization of dienes. Butadiene polymerized with these complexes produces high molecular-weight polymers which may be *cis* or *trans*. The structure of $(\pi\text{-}C_4H_7)_2Ni$ could be represented as:—

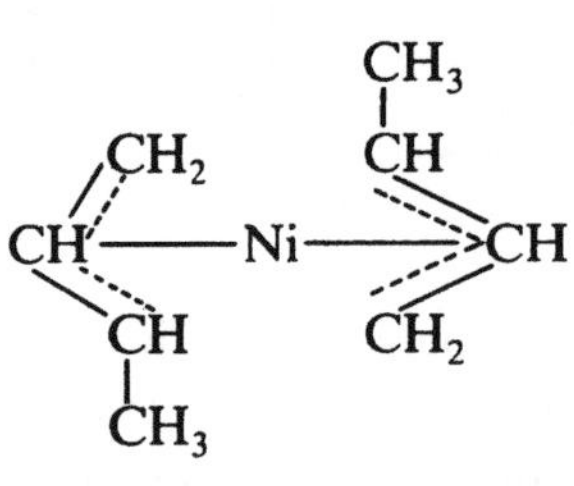

The stereochemistry of the polymerization could well be determined by the addition mode of the diene.

Transition metal halides of titanium, vanadium, molybdenum and tungsten are important coordination catalysts. An organometallic compound (usually from Groups IA and IIIA) is used with the metal halide. An important property of these catalysts is their utility in producing linear unbranched polymers as in linear polyethylene and stereoregular polymers, as in those of propylene, butadiene and isoprene. Ziegler-Natta catalysts are among the most useful stereospecific catalysts. The catalyst system was discovered when ethylene was polymerized using triethylaluminum cocatalyzed with titanium tetrachloride. The polymer formed was found to have higher crystallinity than polyethylenes produced by free radical initiators. The polymerization reaction involves the insertion of an olefin molecule into the bond between the metal and the growing alkyl group. This could simply be represented as:—

$$MR + n\,CH_2{=}CH_2 \rightarrow M{-}(CH_2{-}CH_2)_n{-}R$$

When the same catalyst was used with propylene, isotactic polypropylene was the major product. Propylene polymerized with free radical and some ionic initiators can produce a mixture of three different stereo forms (Fig. 9-1).

a — Atactic: An irregular form where the methyl groups are randomly distributed.
b — Isotactic: A stereoregular form where all methyl groups are found on one side of the polymer chain:
c — Syndiotactic: Another stereoregular form but the methyl groups alternate regularly from one side to the other;

Stereospecific polymerization reactions are those that predominantly produce a

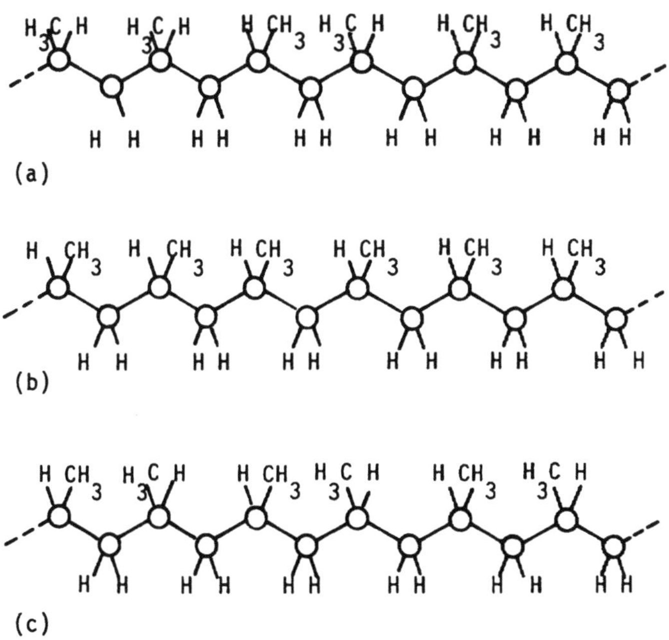

Fig. 9-1. Tactic forms of polypropylene: (a) atactic (b) isotactic (c) syndiotactic.

specific polymeric isomer. The formation of stereoregular polymers with Ziegler catalysts, led to the conclusion that a special type of insertion must be taking place, a *cis*-insertion reaction. A reaction sequence has been suggested to account for the observed polymerization results.

The following is some of the basic information that led Cossee and Arlman to suggest a mechanism to account for the stereospecific nature of Ziegler-Natta catalyst systems:—

a — The catalytic action takes place by the low valency titanium (+3 oxidation state) and in particular α-$TiCl_3$

b — An exchange reaction takes place between triethylaluminum and titanium chloride which could be represented by equation (I)

$$TiCl_3 + Al(C_2H_5)_3 \rightarrow TiC_2H_5Cl_2 + Al(C_2H_5)_2Cl \qquad (I)$$

c — Polymerization of an olefin such as ethylene or propylene takes place in catalytic sites on the surface of the catalyst. This conclusion resulted from the exchange experiments between solid $TiCl_3$ and radioactive $Al(^{14}C_2H_5)_3$.

It was found that the number of radioactive ethyl groups is much lower than is represented by the stoichiometric equation (I). This result was taken to indicate that the exchange reaction must have taken place on the catalyst surface. When the radioactive catalyst was used to polymerize propylene, each polymer molecule was found to have only one radioactive ethyl group

d — It was suggested, then, that the catalyst surface is a five-coordinate compound of a titanium(III) metal which is octahedrally surrounded with five Cl anions and a vacant catalytic site. An exchange with $Al(C_2H_5)_3$ gives the active catalyst.

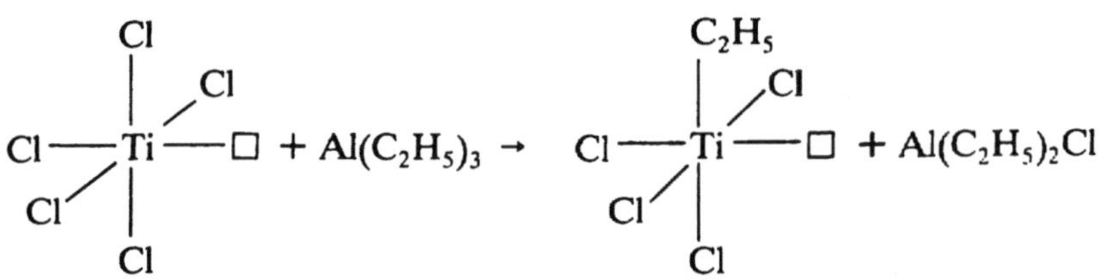

e — When a monomer such as ethylene is used, it occupies the vacant site forming a complex with titanium as shown in equation (II)

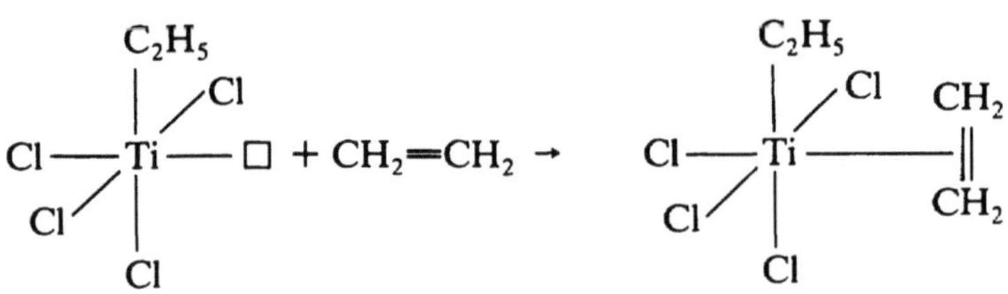

(II)

f — Insertion takes place when the ethyl group becomes detached from the catalyst surface and becomes bonded to the nearest ligand-carbon leaving a vacant site. The other carbon of the ethylene ligand gets attached to titanium as shown in (III).

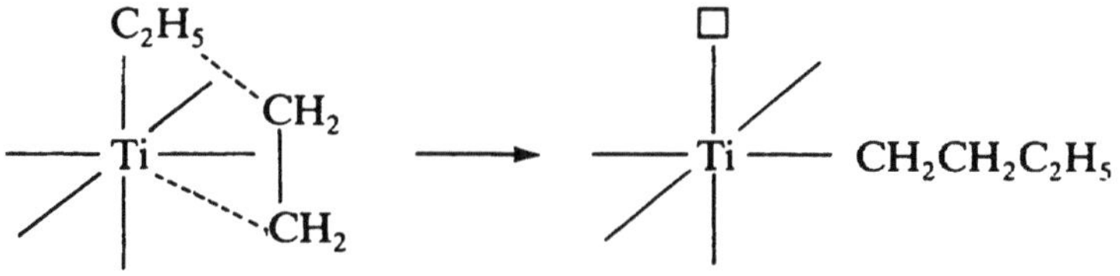

(III)

g — When a second ethylene molecule approaches the vacant site, it becomes attached to it and a second insertion takes place as shown in (IV). This process can continue until termination takes place.

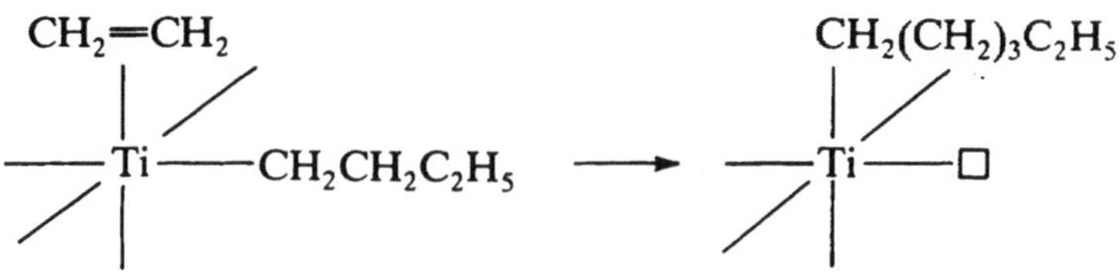

(IV)

However, if this process takes place when polymerizing propylene a syndiotactic rather than an isotactic polymer will be formed, as shown in scheme (a) Fig. 9-2.

As mentioned before, using Ziegler-Natta catalysts for polymerizing propylene produces isotactic polypropylene. The stereospecific nature of these catalysts has been explained by the Cosse-Arlman model which provides information regarding the crystalline nature of α-$TiCl_3$ as a hexagonal close packing. This structure has anion vacancies which allows the growing polymer chains and the attached ligand to be close enough for *cis* insertion to take place. However, due to the difference in steric requirements, one of the two positions available for the ligand to be attached to Ti(III)) must have a greater affinity to accomodate the growing polymer than the other.

In light of this understanding, the propagating polymer then formed returns rapidly back to that site. The formation of isotactic polypropylene is represented in scheme (b) of Fig. 9-2.

Termination of the polymer formed can take place by adding hydrogen to the system.

Another mode of termination is by a chain transfer reaction which can take place in presence of metal alkyls such as $Al(C_2H_5)_3$.

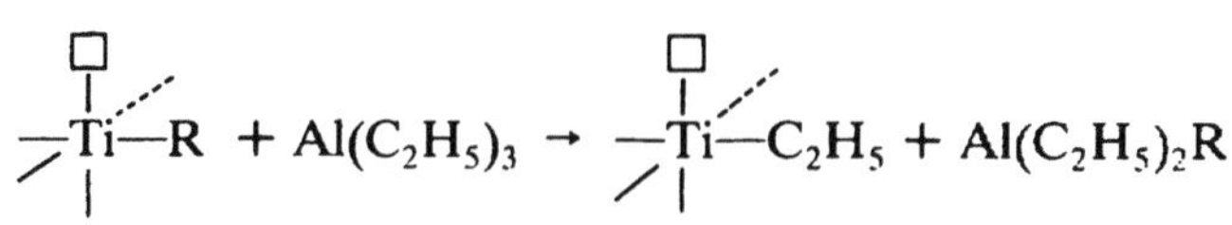

R = polymer chain attached to the transition metal and is exchanged with the ethyl group.

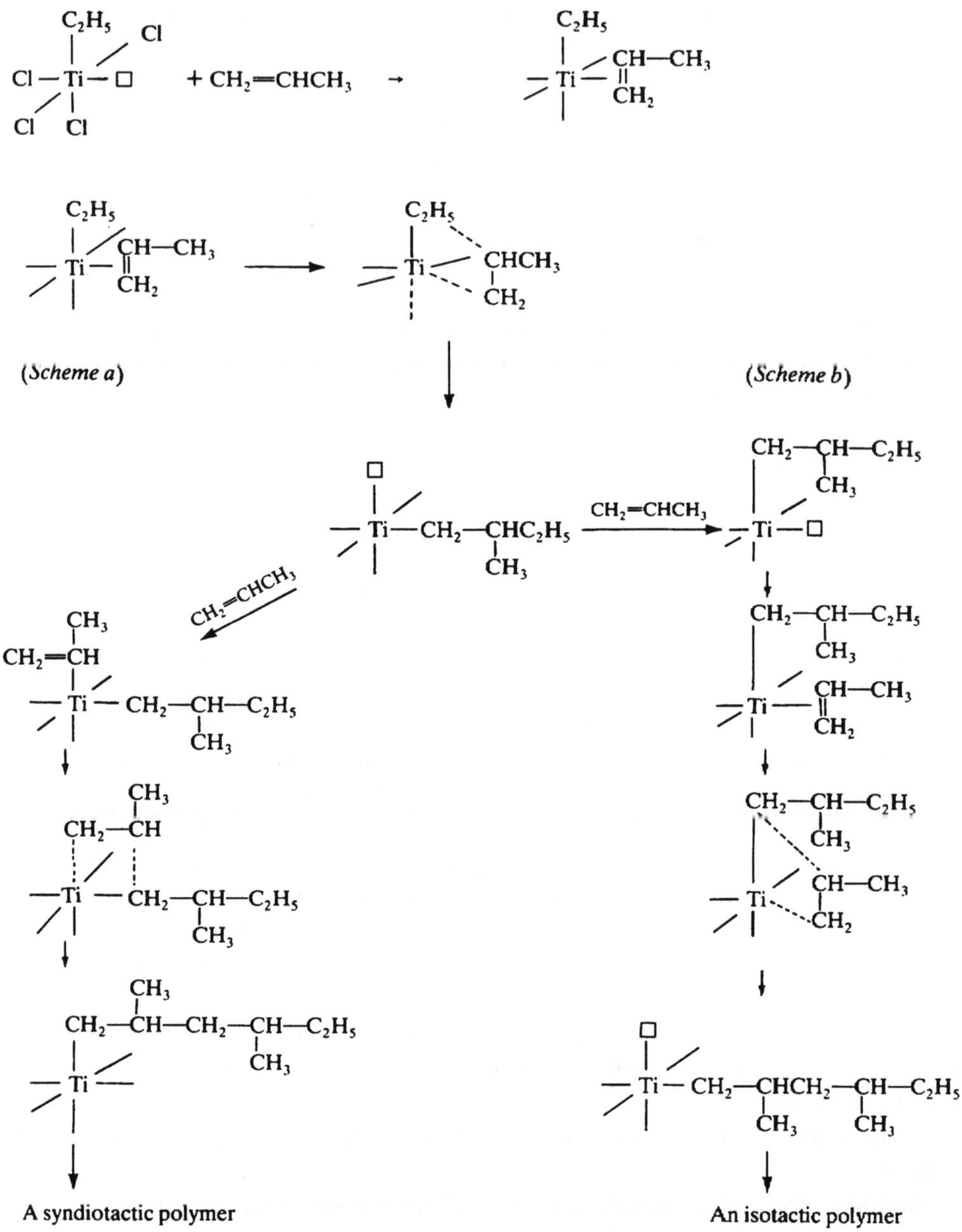

Fig. 9-2. Possible mechanism for the polymerization of propylene to the isotactic form.

The polymer can be recovered from the reaction medium by reacting with alcohol or water.

$$Al(C_2H_5)_2R + R'OH \rightarrow RH + Al(C_2H_5)_2OR'$$

Termination may also take place if the concentration of the monomer is high compared to the concentration of the aluminum alkyl. In case of polymerizing propylene, for example, a chain transfer reaction takes place between the propagating polymer and propylene producing an unsaturated polymer.

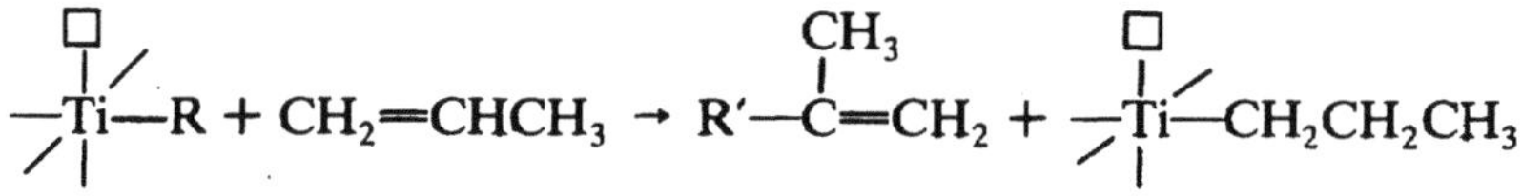

Ring Opening Polymerization. This polymerization does not differ appreciably from the polymerizations mentioned previously. One difference relating to the monomer's structure is that it has to be a cyclic compound that can be opened by the aid of a catalyst during polymerization. Another difference is that polymers formed by a condensation-like reaction do not result in the loss of a small molecule. For example, polymerizing caprolactam to nylon 6 does not produce water, as in the case of polymerizing 11-aminoundecanoic acid.

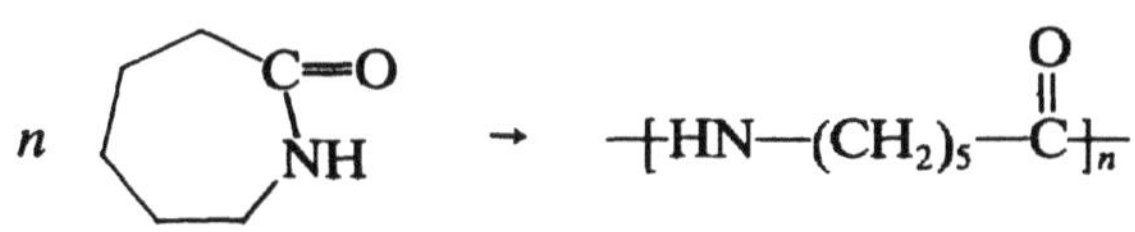

$$n\,H_2N{-}(CH_2)_{10}{-}\overset{\overset{\large O}{\|}}{C}{-}OH \rightarrow {+}HN{-}(CH_2)_{10}{-}\overset{\overset{\large O}{\|}}{C}{+}_n + n\,H_2O$$

Another difference worth mentioning is that, when a cyclic olefin is polymerized by addition, the double bond is retained. For example, polymerizing cyclopentene produces an elastomeric material similar to polymers produced from conjugated diolefins in having a residual double bond in the backbone of the chain.

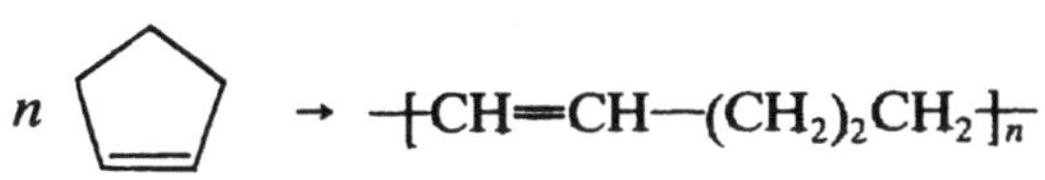

Monomers for ring opening polymerization differ appreciably in their chemical composition. For example, cyclic monomers suitable for polymerization by addition are those possessing one functional group type, such as epoxides, cyclic imines, cyclic anhydrides and cyclic olefins. A driving force for opening small rings, such as ethylene and propylene oxides, is the release of the ring strain. Cyclic monomers for polymerization by ring opening condensation possess two functional groups within the ring. Examples of monomers suitable for this type of polymerization are lactones and lactams. Lactones produce polyesters and lactams produce polyamides when polymerized.

Ring opening polymerizations are either catalyzed by strong bases such as sodium alkoxides and metal hydrides or by protonic acids.

Many important polymers are produced by ring opening. For example, ethylene oxide, is polymerized to a high molecular weight polymer in presence of methoxide ions.

$$CH_3O^-Na^+ + \overset{O}{\overbrace{CH_2—CH_2}} \rightarrow CH_3OCH_2CH_2O^-Na^+ \qquad \text{Initiation}$$

$$CH_3OCH_2CH_2O^-Na^+ + n\ \overset{O}{\overbrace{CH_2—CH_2}}$$
$$\rightarrow CH_3[OCH_2CH_2]_nOCH_2CH_2O^-Na^+ \qquad \text{Propagation}$$

$$CII_3[OCII_2CII_2]_nCII_2CII_2O^-Na^+ + CH_3OH$$
$$\rightarrow CH_3[OCH_2CH_2]_nCH_2CH_2OH + NaOCH_3 \qquad \text{Termination}$$

Epoxy resins are important general-purpose polymers made by a base catalyzed reaction between epichlorohydrin and a polyhydroxy compound such as bisphenol A.

$$\overset{O}{\overbrace{CH_2—CH}}—CH_2Cl + HOROH \rightarrow HOROCH_2\overset{OH}{\overset{|}{C}}HCH_2Cl$$

$$HOROCH_2\overset{OH}{\overset{|}{C}}HCH_2Cl + NaOH \rightarrow HOROCH_2\overset{O}{\overbrace{CH—CH_2}} + NaCl + H_2O$$

HOROH =

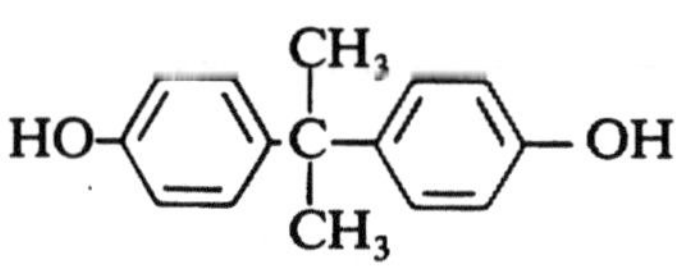

bisphenol A

The linear polymer may then be cured by the use of an acid anhydride that reacts with the hydroxyl groups or through the terminal epoxide likage by the use of an amine.

9.4. Polymerization Techniques

Addition polymerization reactions are exothermic because they are essentially bond forming. Therefore it is necessary to control the heat of the reaction. This can be done to a certain extent by the use of an appropriate solvent, or by limiting the reaction to a low conversion and by the use of external cooling.

Another feature of addition polymerization which requires control is the progressive increase in viscosity due to polymer growth. This is solved, as with the heat effect, by the use of a solvent or by carrying the reaction to a low conversion. The main polymerization techniques are; bulk, solution, suspension and emulsion. A special technique is sometimes used to polymerize reactive compounds is termed interfacial polymerization.

In bulk polymerization, no solvent is used and impurities in the final polymer are only those that are found in the original monomer. Heat and viscosity controls are important because of the absence of a solvent.

Solution polymerization, is achieved by dissolving the monomer in an organic solvent. Heat and viscosity controls are easier than in bulk polymerization because of the presence of the solvent. In certain applications such as in adhesives and coatings, removal of the solvent may not be necessary.

In suspension polymerization, the monomer is dispersed in a liquid such as water by the use of mechanical agitation to keep the monomer dispersed in the liquid. Stabilizers such as talc, polyvinyl alcohol etc. are used. Emulsion polymerization requires the use an emulsifying agent. Emulsifiers may be ionic, such as benzene sulfonates, or nonionic, such as polyvinyl alcohol.

Interfacial polymerization is used when acid chlorides are condensed with diols or diamines to produce polyesters or polyamides. The reaction is fast and takes place at the interface and the polymer is continuously removed from the surface.

9.5. Some Important Polymers Formed by Coordination Catalysts

Many commercially important polymers are produced using coordination compounds and Ziegler type catalysts. Among these are linear alcohols and α-olefins for detergents, linear low density polyethylene (LLDPE), high density polyethylene and isotactic polypropylene for plastics, polybutadiene, polyisoprene and *trans*-polypentamer for rubber. A brief discussion of the production of these polymers is given below.

9.5.1. PRODUCTION OF 1-BUTENE

Ethylene may be dimerized to 1-butene by the use of conventional dimerization catalysts such as NiH. These catalysts produce a high yield of 1-butene due to the high activity of the catalyst for hydrogen transfer reactions.

$$NiH + CH_2{=}CH_2 \rightarrow NiCH_2CH_3$$

$$NiCH_2CH_3 + CH_2{=}CH_2 \rightarrow NiCH_2CH_2CH_2CH_3$$

$$2\ Ni(CH_2)_3{-}CH_3 \xrightarrow{\text{H transfer}} \underset{\text{Main product}}{CH_2{=}CHCH_2CH_3} + CH_3CH{=}CHCH_3 + 2\ NiH$$

However, oligomers of ethylene are also co-produced from this reaction due to polymerization of ethylene.

$$Ni(CH_2)_3{-}CH_3 + CH_2{=}CH_2 \rightarrow Ni(CH_2)_4{-}CH_2CH_3 \xrightarrow{CH_2{=}CH_2} \text{Oligomers}$$

A new titanium(IV)-based catalyst is highly selective for the dimerization reaction.

This is probably due to the absence of hydrogen which is responsible for the isomerization of 2-butene to 1-butene, as is the case when conventional catalysts are used. A cyclic intermediate has been proposed for Ti(IV) catalytic action.

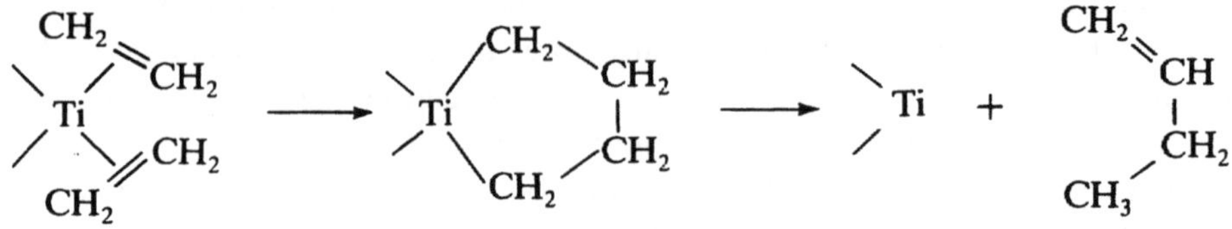

A modifying agent is used to prevent the Ti(IV) complex reverting to the Ti(III) complex, responsible for producing high molecular weight polymers.

1-Butene is used for the production of some chemicals such as methyl ethyl ketone and butylene oxide. It is also used as a comonomer for the production of linear low density polyethylene.

9.5.2. PRODUCTION OF LINEAR ALCOHOLS

Linear alcohols (C-12—C-18) are used for the production of biodegradable detergents. They are produced by an insertion reaction using a Ziegler-Natta catalyst of α-titanium trichloride and triethyl aluminum. The alkyl aluminum is prepared by reacting aluminum with ethylene and hydrogen.

$$6\ CH_2{=}CH_2 + 3\ H_2 + 2\ Al \rightarrow 2\ (CH_3CH_2)_3Al$$

The polymerization reaction takes place under high pressure and can be represented as follows:

$$1\tfrac{1}{2}n\ CH_2{=}CH_2 + (CH_3CH_2)_3Al \rightarrow [CH_3{-}(CH_2)_{n+1}]_3Al$$

The trialkyl aluminum produced is further oxidized to give the trialkoxide.

$$[CH_3(CH_2)_{n+1}]_3Al + 1\tfrac{1}{2}\ O_2 \rightarrow [CH_3(CH_2)_{n+1}O]_3Al$$

The last step is the acid hydrolysis of the oxides to give a mixture of even numbered alcohols and aluminum sulfate.

9.5.3. PRODUCTION OF α-OLEFINS

α-Olefins in the C-12—C-14 range are used for the alkylation of benzene for the production of ionic detergents. α-Olefins are produced by dehydrogenation of the corresponding n-paraffins, or by oligomerization of ethylene using Ziegler catalysts.

$$(CH_3CH_2)_3Al + 1\tfrac{1}{2}n\ CH_2{=}CH_2 \rightarrow [CH_3(CH_2)_{n+1}]_3Al$$

$$[CH_3(CH_2)_n]_3Al + 3\ CH_3CH_2CH{=}CH_2 \rightarrow 3\ CH_3(CH_2)_{n-2}CH{=}CH_2 + (CH_3CH_2CH_2CH_2)_3Al \quad n > 3$$

Currently, alkylation of benzene with α-olefins is often catalyzed with HF while alkylation with monochlorinated paraffins utilizes a Friedel-Crafts catalyst. The alkylate is then sulfonated and neutralized.

9.5.4. PRODUCTION OF LINEAR POLYETHYLENES

Polyethylene is the most extensively used thermoplastic. It is characterized by being a flexible material which is resistant to many chemicals. Polyethylene produced by free radical initiators is branched and is more flexible due to a lower crystallinity than linear high density polyethylene. A new grade of low density polyethylene, which is linear, is produced by the use of Ziegler catalysts. The polymer may be copolymerized with 1-butene to improve its properties.

Table 9-2 compares the properties of high and low density polyethylene.

High density polyethylene is produced by a low pressure process in a fluidized bed reactor. Either Ziegler type catalyst or a mixture of metal oxides such as chromium oxide or molybdenum oxide on alumina or silica—alumina carrier is used. Both catalyst systems are selective for producing a highly crystalline linear polymer.

Ethylene polymerization is highly exothermic, and heat control is very important. Heat removal is usually done by circulating the ethylene gas through heat exchangers, then compressing it for recycling to the reactor. Either a liquid or a gas phase process is used. In the liquid phase process a hydrocarbon diluent is added which requires a special recovery system. Some branching for HDPE is generally incorporated in the backbone of the polymer by adding a comonomer such as hexene. Molecular weight control can be obtained by introducing hydrogen to the circulating gas. Typical reaction conditions for HDPE processes are approximately 100 °C and 10—25 bar.

As a thermoplastic, polyethylene can be molded into almost any shape, extruded into fiber or filament, and blow molded. About 85% of all blow-molded bottles are produced from HDPE.

Pipes from HDPE are flexible, tough and corrosion resistant. They are extensively used for irrigation and for carrying corrosive chemical materials.

Table 9-2. Typical properties of polyethylene.

Polymer	Melting Point range °C	Density	Degree of crystallinity %	Stiffness modules psi × 10^3
Branched, Low density	107—121	0.92	60—65	25—30
Medium density		0.935	75	60—65
Linear High Density				
Ziegler type	125—132	0.95	85	90—110
Phillips type		0.96	91	130—150

9.5.5. PRODUCTION OF POLYPROPYLENE

Isotactic polypropylene is produced by the use of a Ziegler type catalyst. The stereoregular polymer is highly crystalline. Polymerization may be carried in the gas phase, in bulk or in a diluent slurry process. In the gas phase process a fluidized bed reactor is used. A reaction temperature range of 50—100°C and a pressure range of 15—30 bar are typical. In this process, agitation is provided by the flow rate of the fluidizing gas. Mechanically agitated gas reactors may also be used to ensure good mixing of the catalyst and heat removal is ensured by condensing the refluxing vapors. New processes use catalysts which are highly active and selective. In these processes there is no need for a catalyst removal system since the amount of the catalyst in the polymer is negligible (4 ppm versus 300 ppm in older processes). Polymers produced have an isotactic index of 93 or higher (isotactic index is expressed as the weight % of the isotactic form divided by the total weight of the produced polymer).

A new catalyst system made of three components — titanium supported on magnesium halide, an organometallic component and a Lewis base — is claimed to produce a polymer having an isotactic index of 99. In older processes the product polymer is washed with water or alcohol to decompose the catalyst. Solvent recovery is achieved by centrifugation—filtration or by steam followed by hot air. The polymer is obtained in the form of a powder which can be processed to pellets.

Polypropylene, as a thermoplastic material, possesses favorable properties such as light weight (lightest thermoplastic), good chemical, heat and abrasion resistance, low water absorption, and relatively low cost. Table 9-3 shows the important physical properties of polypropylene. Copolymers of propylene with ethylene have elastomer-like properties and are called thermoplastic elastomers. Products are used for seals, tubing, shoe soles etc. Elastomers of this type do not possess the double bonds necessary for crosslinking. To provide some residual double bonds, a third monomer is used, such as 1,4-hexadiene or ethylidine norbornene. Cross-linking may also be achieved by using peroxides without using a third monomer. About 30% of polypropylene is used for fiber production. Low cost fibers made from polypropylene are replacing those made from sisal and jute. Polypropylene polymers have a low softening point, which effects their processing by melt spinning techniques. The softening point could be improved by copolymerizing

Table 9-3. Properties of polypropylene.

Density g/cm^3	0.9—0.91
Specific volume cm^3/lb	30.4—30.8
Tensile strength psi	3200.0—5000.0
Water absorption, 24 hr., %	0.01
Elongation, %	3.0—700
Melting point, T_m °C	176

with monomers having polar groups such as methyl methacrylates. Polypropylene fibers are used in tufted carpets, upholstery fabrics and underwear blends.

9.5.6. POLYBUTADIENE

Butadiene can be polymerized using different initiator types. When free radicals are used the addition is predominantly a 1,4- along with a small amount of the 1,2-isomer. The 1,4-addition is not stereoselective but a mixture of *cis*- and *trans*-1,4-addition products is obtained.

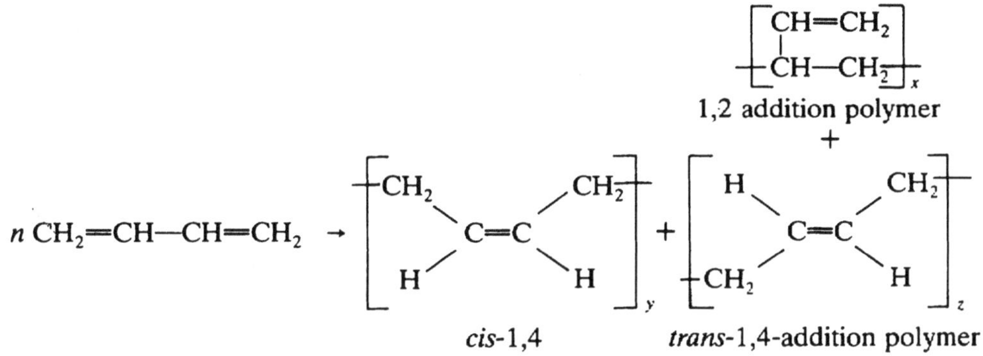

Polymerization of butadiene using stereospecific catalysts produces predominantly either *cis*- or *trans*-1,4-addition polymers. Alfin catalysts, which are composed of an alkenyl sodium, an alkoxide and alkali metal halide, produce mainly a *trans*-1,4-polymer.

Anionic catalysts such as lithium and alkyl lithium produce a predominantly 1,4-addition polymer but with different levels of *cis*- and *trans*-isomers. A high *cis*-1,4-addition polymer is produced, however, by the use of coordination and Ziegler-type catalysts.

Butadiene is readily copolymerized with other monomers. When copolymerized with styrene, SBR is obtained. SBR is a well-known rubber for general tire production. Butadiene is also copolymerized with acrylonitrile and the product is ABR. The most important use of polybutadiene (*ca.* 95%) is in the tire industry. This is a result of its favorable characteristics such as high abrasion resistance, high resilience and its ability to form a blend with natural rubber.

9.5.7. POLYISOPRENE

As with butadiene, polymerization of isoprene gives a mixture of isomers if the conditions are not controlled for stereospecific addition. Since the isoprene molecule is not symmetrical, three types of addition reactions are possible; 1,2-,

1,4- and 3,4-. Six isotactic forms are possible from 1,2- and 3,4-addition and two geometrical isomers (*cis* and *trans*) from 1,4-addition.

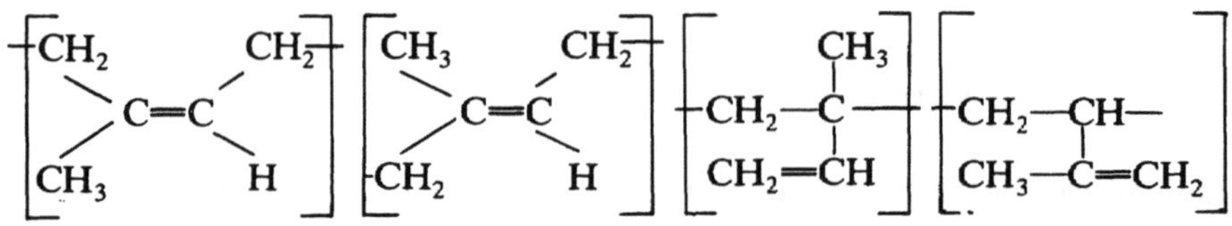

Commercial polymerization of isoprene produces mainly a mixture of the 1,4-addition products. Either lithium compounds or Ziegler type catalysts are used to selectively produce the *cis*-1,4-addition product. The polymerization rate and the *cis*-level depend on the Al/Ti ratio when using Ziegler catalysts. The structure of *cis*-1,4-polyisoprene is similar to natural rubber, which possesses a high degree of elasticity due to weak intermolecular forces. The *trans*-1,4-isomer is highly crystalline and is therefore not elastic. Crosslinking of the isoprene polymer takes place by heating with sulfur. This process is called vulcanization where sulfur bridges are formed due to presence of reactive allylic positions.

Isoprene (2.5%) is copolymerized with isobutylene (97.5%) to produce butyl rubber. Butyl rubber is mainly used for the production of tire inner tubes because of its low permeability to air. It is also used for cable covers, mechanical goods and adhesives.

9.5.8. TRANS-POLYPENTAMER

The catalyzed ring opening polymerization of cyclopentene produces a polymer which has elastomeric properties. Polymerization in solution using tungsten hexachloride and an aluminum alkyl gives predominantly the *trans*-isomer.

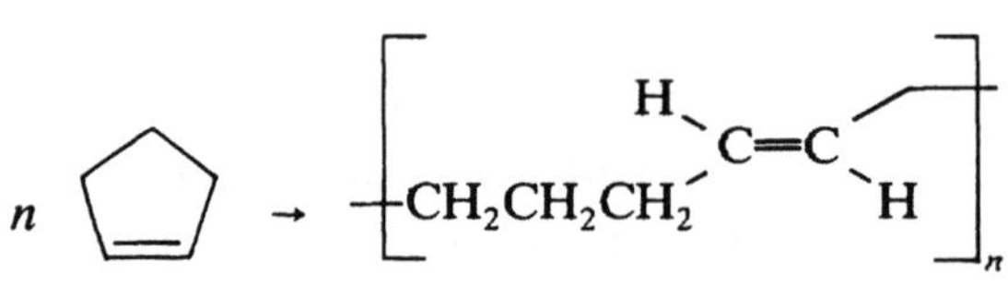

The polymer is mainly amorphous with a linear structure. It can be vulcanized since the residual double bonds are retained. However, a drawback of using it for tire production is its low skid resistance.

9.5.9. POLYCHLOROPRENE

Anionic polymerization of chloroprene produces a highly *trans*-1,4-polychloroprene

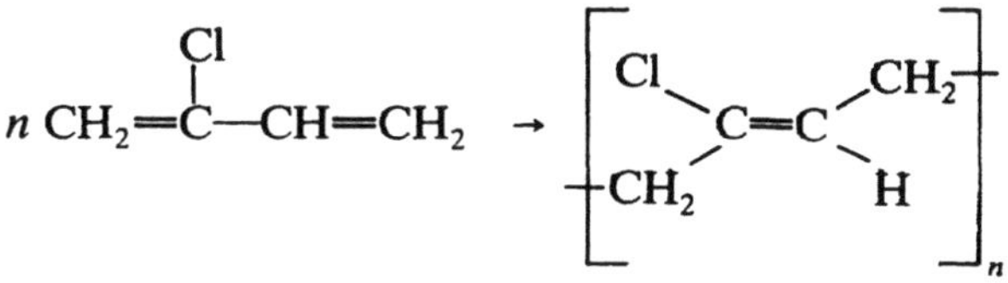

Polychloroprene is used under extreme service conditions because of its high resistance to oil and chemicals. Conveyor belts, cables, adhesives are major outlets for polychloroprene.

References

Allcock, H. and Lampe, F.: *Contemporary Polymer Chemistry*, Prentice-Hall, Inc. Englewood Cliffs, New Jersey, 1981.

Arlman, E. and Cosse, P. J.: *Catal.* **3**, 99 (1964).

Commereue, D., Chauvin, Y., Gaillard, J., and Leonard, J.: 'Dimerize Ethylene to Butene-1', *Hydrocarbon Processing*, Nov. 1984, p. 118—120.

DiDrusco, G. and Rinaldi, R.: 'Polypropylene-Process, Selection Criteria', *Hydrocarbon Processing*, Nov. 1984, pp. 113—117.

Gates, B., Katzer, J., and Schuit, G.: *Chemistry of Catalytic Processes*, McGraw-Hill Book Co., New York, 1979.

Hatch, L. and Matar, S.: *From Hydrocarbons to Petrochemicals*, Gulf Publishing Co. Houston, Texas, 1981.

Keii, T.: *Kinetics of Ziegler Natta Polymerization*, Kondansha Ltd. Tokyo 1981, pp. 64—68.

Natta, G. and Danusso, F.: *Stereoregular Polymers and Stereospecific Polymerizations*, Pergamon, Oxford, 1967.

Natta, G., Porrie, A., and Stoppa, G. J.: *Makromol Chem.* **77**, p. 144 (1964).

Natta, G. J.: *Polymer Science* **48**, p. 219 (1960).

Ravve, A.: *Organic Chemistry of Macromolecules*, New York, Marcel Dekker, Inc., 1967.

CHAPTER 10

Experimental Techniques in Catalysis

10.1. Introduction

The suitability of chemical compounds as catalysts for industrial processes is determined mainly by the following characteristics:

- activity
- selectivity
- lifetime (= stability)

The *activity* of a catalyst is a quantitative measure of the ability of a compound to catalyze a certain chemical reaction. It is equivalent to the reaction rate in uncatalyzed reactions. A variety of activity related parameters is used in the

Table 10-1. Parameters describing the quality of a catalyst.

(a) activity a (sometimes called reaction rate **r**): $a = \frac{\text{converted reactant}}{\text{mass of catalyst} \times \text{time}}$	$\frac{\text{mol}}{\text{kg} \times \text{h}}$
(b) gas hourly space velocity $GHSV$: $GHSV = \frac{\text{volume of gas passed through the reactor}}{\text{catalyst volume} \times \text{time}}$	$\frac{\text{L}}{\text{L} \times \text{h}}$
(c) liquid hourly space velocity $LHSV$: $LHSV = \frac{\text{volume of liquid passed through the reactor}}{\text{catalyst volume} \times \text{time}}$	$\frac{\text{L}}{\text{L} \times \text{h}}$
(d) yield A (as %): $A = \frac{\text{desired product}}{\text{reactant fed}} \times 100$	$\frac{\text{mol} \times 100}{\text{mol}}$
(e) conversion U (as %): $U = \frac{\text{converted reactant}}{\text{reactant fed}} \times 100$	$\frac{\text{mol} \times 100}{\text{mol}}$
(f) selectivity S (as %): $S = \frac{\text{mole of product } i}{\text{mole of all products}} \times 100$	$\frac{\text{mol} \times 100}{\text{mol}}$
(g) space time yield STY: $STY = \frac{\text{mole (or kg) of desired product}}{\text{volume of catalyst} \times \text{time}}$	$\frac{\text{mol (kg)}}{\text{L} \times \text{h}}$
(h) catalyst lifetime Z: $Z = \frac{\text{converted reactant before deactivation of the catalyst}}{\text{mass of catalyst} \times \text{time}}$	$\frac{\text{kg}}{\text{kg} \times \text{h}}$
(i) residence time t (= reciprocal value of the $GHSV$): $t = \frac{\text{catalyst volume} \times \text{time}}{\text{volume of gas passed through the reactor}}$	$\frac{\text{L} \times \text{h}}{\text{L}}$

literature. Some of them are compiled in Table 10-1. The activity of a catalyst depends not only on its chemical nature but also on the area of the active surface, the pore size distribution, and the presence of modifiers, promotors, or inhibitors. These substances are usually present in small concentrations. They change the properties of the catalyst, but they are not catalytically active alone.

The *selectivity* describes the ability of a catalyst to favor one reaction pathway over all other possibilities. A high selectivity for the desired product is a very important criterion for the quality of a catalyst. By-products are often a nuisance which must be disposed of by incineration or other methods.

The *lifetime* of a catalyst is the time which a catalyst can stay in operation before it must be replaced because of loss of activity. This time is determined by the mechanical strength of the catalyst and its resistance to poisoning. A catalyst poison is a substance which lowers the activity of a catalyst. Typical poisons are sulfur compounds (e.g., H_2S) or heavy metals (e.g., lead).

Three important steps precede the application of a catalyst in an industrial reaction. It must be prepared, characterized, and tested. This is relatively easy for homogeneous catalysts, since conventional chemical and analytical methods can be used. For heterogeneous catalysts, however, the preparation and characterization may look like black magic. Therefore the basic methods used in heterogeneous catalysis are discussed in some detail in this chapter.

10.2. Preparation of Heterogeneous Catalysts

Heterogeneous catalysts may consist of metals, metal oxides, metal salts, or nonmetal oxides. Very often the active catalyst is supported by another material. The support may be silica, alumina, active carbon, etc. Its function is threefold:

(1) It dilutes the active catalyst and makes it more accessible to the reactants. Very often a support reduces the amount of active component which is required to achieve a certain activity. This is obviously an advantage when expensive compounds or noble metals are used as active catalysts.
(2) A support can enhance the mechanical stability of a catalyst.
(3) The support can participate in the reaction by activating the reactants or by favoring certain reaction pathways over others.

Often, all three effects are utilized in industrial processes.

Catalysts can be made in many different macroscopic forms. They can be applied as a powder, a sand-like granulate, tablets or pellets. Automobile exhaust gas or flue gas denitrification catalysts are usually monolites. These are porous blocks with channels parallel to the gas flow. They allow very high flow rates, since they cause a small pressure drop along the catalyst.

A skeleton catalyst consists of a metal skeleton, obtained by dissolving one component of an alloy with acid or base leaving the more noble metal component. This type is also known as a Raney catalyst.

Heterogeneous catalysts are chiefly produced by one of the four general processes: mixing, impregnation, precipitation, or metal alloy treatment by partial corrosion (Raney catalysts).

10.2.1. MIXING

The components of a catalyst are mixed together as solid or — more often — as a solution of a metal salt with a suspended solid support. In the latter case the solution is drawn into the solid by capillary forces. After filtration the solid powder is dried and then pelletized, extruded, or tableted. Often lubricants or binders are added to enhance the mechanical properties of the pellets or tablets. The pressure applied in this step can influence the properties of the final catalyst profoundly. The higher the pressure the smaller is the average pore size of the catalyst.

10.2.2. IMPREGNATION

The impregnation method is similar to the mixing process. Here, however, the support is pressed into tablets before it is impregnated with a solution of the metal salt. The metal ions diffuse into the support and form a concentration profile in which more ions are located near the surface than in the interior of the pellet. In contrast, the metal distribution is more uniform, than when the catalyst is prepared by the mixing method described above.

10.2.3. PRECIPITATION

This method is often used to prepare oxidic or sulfidic catalysts. A solution of a metal salt (often the nitrate) is combined with the ammonium salt of a suitable anion to precipitate the metal ion as hydroxide, sulfide, oxalate, carbonate, etc. Upon heating to high temperatures (calcination) the anions are decomposed to give the metal oxides. Simultaneous addition of the metal salt and the precipitating agent to the precipitation flask gives a more uniform catalyst composition than sequential addition of the components.

10.2.4. SKELETON CATALYSTS

Skeleton or Raney catalysts are prepared by partial corrosion of a suitable alloy. The metal alloy is treated with a reactant, for example an acid or a base, which dissolves or corrodes one component of the alloy leaving the others in a form with a large, highly active surface.

Examples are the preparation of highly active copper by dissolving zinc from Zn-Cu alloys with acid and the preparation of Raney nickel from a 1/1 Ni-Al alloy by dissolving the aluminum with base.

The actual preparation of the catalyst by mixing, impregnation, or precipitation is followed by a sequence of steps, which are also crucial for the catalyst properties: washing, drying, calcination, and activation.

Washing. The raw catalyst is washed with an excess of water to remove any traces of the precipitating agent or the excess of the impregnating agent. Sometimes additives are added to the washing water (e.g., ammonia) to enhance its effectiveness.

Drying is the most tricky procedure, since it can change the properties of a catalyst drastically. As a general rule, the catalyst loses activity, stability, surface area, uniformness, etc., when dried at too high a temperature or too rapidly. Therefore the catalyst is dried with moist air under carefully controlled conditions. Special care must be taken, when a laboratory procedure is transferred to the manufacturing stage, since the drying rates change upon scale-up.

The drying takes place in two distinct steps:

1. During constant rate drying, liquid is evaporated from the external surface of the pellets or tablets. The drying rate is determined by the rate of water vapor diffusion through the interfacial gas film around each pellet. The water vapor is then carried away by the large volume of air flowing through the catalyst. This type of drying is not affected by the properties of the solid, but mainly by the temperature, air flow rate, and air humidity.
2. After the external moisture is evaporated the second step the 'falling rate drying' starts. Here water moves from inside the pellets to the surface. The rate of this process is determined by pore diffusion, capillary forces, surface migration, or other properties of the solid. The drying rate decreases with decreasing moisture content until the catalyst is dry. The temperatures for drying are usually between 60 and 100°C.

Calcination. Operationally calcination and drying differ only in temperature. Upon calcination at 200 to 1200°C the chemical and physical properties of the catalyst change: The catalyst support may undergo phase transitions, the pore diameter and the pore size distribution may change, lubricants, binding agents, and anions are removed, partial sintering increases the mechanical strength, metals are oxidized, physical mixtures of oxides crystallize to form mixed oxides, and the carrier may react with the active catalyst.

Again, special care must be taken when a laboratory procedure is scaled up to the manufacturing stage, since the temperature increases more rapidly in a laboratory experiment than in a large-scale apparatus. This may lead to quite different properties of a catalyst which is intended for large scale use.

Activation. Sometimes it is necessary to activate a catalyst before it is actually put into operation. This is usually the case when the catalyst is active in a reduced oxidation state. In such a case the catalyst is transferred to the reactor and treated with hydrogen or another reducing agent before it is brought into contact with the reactants.

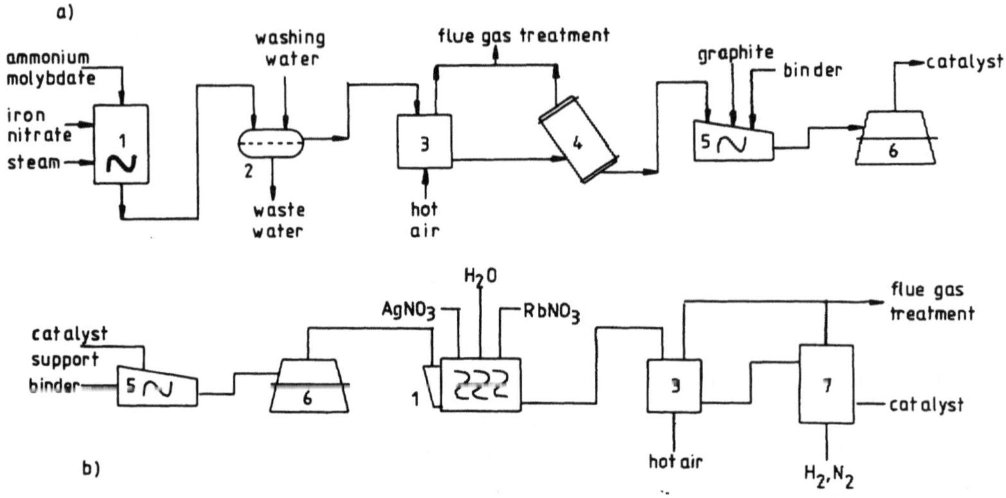

Fig. 10-1. Flow diagram of typical catalyst preparation processes. (a) Precipitation: The example is the preparation of the iron molybdate catalyst used for the production of formaldehyde from methanol. (b) Impregnation: The example is the preparation of the silver catalyst for the epoxidation of ethylene. 1 = reactor; 2 = filtration unit; 3 = dryer; 4 = calcination furnace; 5 = dry mixing; 6 = pellet formation; 7 = reactor for the reduction (activation) of the catalyst.

10.3. Characterization of Heterogeneous Catalysts

Composition. The chemical composition of a catalyst is determined by an elemental analysis. This can be done by a classical wet chemical analysis or by more modern methods, like AAS (= Atomic Absorption Spectroscopy), X-ray fluorescence, microprobe, etc. It is important to determine the water content of a catalyst, before any other analysis is performed. The simplest method is to measure the weight loss upon heating. Trace elements can change the properties of a catalyst profoundly. Therefore their concentration should be known. X-ray diffraction and electron microscopy provide information on the nature, size, and distribution of the different phases and crystallites in the catalyst.

Structure. The pore size and pore size distribution can be determined by the mercury or gas penetration methods, in which the catalyst is exposed to varying pressures of mercury, helium or other gases and the pressure difference between the catalyst sample and a nonporous substance of the same volume allows the calculation of the pore radii and pore size distribution. The total internal and external surface area of a catalyst is usually determined by measuring the adsorption isotherms of a gas, normally nitrogen, applying the equation of Brunauer, Emmett, and Teller (BET-method).

Thermal analysis. Upon heating the catalyst a number of processes can occur, for example phase changes, desorption of gases, water, or other small molecules, decomposition, etc. These reactions can be studied in the laboratory using differential thermal analysis (DTA) and differential thermal gravimetry (DTG): A sample of the catalyst is heated at a constant rate and the temperature and the weight of the sample are measured simultaneously.

The desorbed molecules may also be analyzed during the controlled heating process by mass spectrometry, gas chromatography, or Fourier transform infrared (FTIR) spectroscopy.

Mechanical strength can be determined by a number of methods. To test the pressure stability a catalyst pellet is placed between two plates and compressed until it breaks. To measure the impact stability of a catalyst, a pellet is dropped from different heights onto a surface until it breaks.

Surface analytical techniques. The reactivity of a catalyst is mainly determined by the properties of its surface. Therefore many techniques have been developed to obtain information on the molecular layers of a surface. The methods are usually named in abbreviations or acronyms, which are difficult to decipher for non-specialists. Some of them are listed below:

AEM = Analytical electron microscopy determines images with a near atomic resolution and measures the elemental composition by a combination of an electron microscope with X-ray fluorescence. (XES = X-ray emission spectroscopy)

AES (= Auger electron spectroscopy) is also used to obtain information on the elemental composition but moreover on the chemical state of a catalyst surface. The resolution is somewhat smaller (10 to 100 nm) than for AEM.

DRIFT = Diffuse reflectance infrared Fourier transform analysis gives information about the molecules adsorbed on the surface of a catalyst.

ESCA = Electron spectroscopy for chemical analysis is used to determine the elemental composition and the chemical state composition of a surface with a micro- or millimeter resolution. The surface is exposed to X-rays which knock off electrons from the surface atoms. XPS = X-ray photoelectron spectroscopy is the same as ESCA.

EXAFS = Extended X-ray absorption fine structure spectroscopy is used to determine chemical states and the distribution of elements with a mm resolution.

ISS = Ion scattering spectroscopy allows an elemental analysis of the outermost atomic layer by exposing a surface to low-energy ions.

NBS = nuclear backscattering spectroscopy uses high-energy ions to determine the elemental composition of a surface with a depth profile of 10 to 20 nm.

SEM = scanning electron microscopy is used to image the surface morphology of a catalyst.

SIMS = Secondary ion mass spectroscopy gives molecular and elemental information by measuring the ions and fragments which are emitted, when a surface is exposed to low energy ions.

UPS = Ultraviolet photoelectron spectroscopy is very similar to ESCA with the difference that UV light is used to stimulate the emission of electrons from a surface.

TEM = Transmission electron microscopy.

LEED = Low-energy electron diffraction is similar to X-ray diffraction and provides information on the lattice structure of a catalyst surface.

10.4. Testing Catalysts in the Laboratory

As mentioned in the introduction to this Chapter, the quality of a catalyst is determined by a number of factors, such as activity, selectivity for a certain product, and stability. These parameters are themselves functions of temperature, pressure, concentration, residence time, reactor dimensions, catalyst form, etc. Therefore quite a number of variables must be tested in the laboratory before a catalyst is ready for large-scale application. The screening is done by comparing a new catalyst with an established catalyst under exactly comparable conditions. Usually series of parallel reactors are used for catalyst testing. Many different test reactor designs have been used so far. They can be classified according to various criteria.

Integral reactors are operated in such a way that most of the starting material is converted into products on one pass through the reactor. That means, the concentrations of the reactants decrease and those of the products increase along the reactor length. Such an integral reactor can work under isothermal or adiabatic conditions. *Isothermal* means, that all the catalyst is kept at a constant temperature by supplying or removing heat for endothermic or exothermic reactions, respectively. When the reactants are preheated to the reaction temperature before they enter the catalyst bed and when no heat is added to or removed from the catalyst, the reactor is in an *adiabatic* operation mode. In this mode the temperature of the reaction mixture changes along the reactor.

In *differential reactors* the reactants pass through a small layer of the catalyst and the conversion is kept low ($<5\%$). In this way the concentration of the reactants remains nearly constant and it easy to determine the kinetic parameters of the reaction. It is however sometimes difficult to determine the product yields accurately at these low conversions.

10.4.1. EXAMPLES OF INTEGRAL REACTORS

The simplest way to test a catalyst is to suspend it in the reactants contained in a stirred flask. This is the situation of an integral, isothermal reactor. It can only be operated batchwise and gives reliable results only when the activity of the catalyst does not change during the reaction.

The most frequently used test reactor is the integral, isothermic fixed bed reactor. It consists of a quartz or stainless steel tube (e.g., length 1 m, inner diameter 5 cm), which has a perforated plate at the lower end. It is filled with the catalyst and placed into an oven. Sometimes some inert material is stuffed into the upper part of the tube on top of the catalyst to heat up the reactants before they enter the catalyst layer. In the simplest case the concentration of the products is measured only at the end of the tube by gas chromatography or other methods. However, it is possible to have additional ports for analysis and temperature measurements along the reactor. Since the composition of the reaction mixture changes along the reactor, the reaction takes place with different rates at different points in the catalyst bed. Therefore the kinetics can be evaluated only when the concentration and temperature profiles along the reactor are known.

A constant temperature can be maintained by means of a hot air furnace or a heat transfer agent (water, oil, molten salt), which is circulated around the reactor.

When a preheated gas stream is fed into an electrically heated fixed bed reactor usually an adiabatic operation mode arises. This sometimes has advantages, since it may allow a better simulation of the industrial process.

In moving bed reactors the catalyst moves slowly through the reactor (most commonly downwards) and the reactants pass through it in the opposite direction. The spend catalyst can then be regenerated offline in a separate procedure. This method has some advantages in applications in which the catalyst is rapidly poisoned.

A fluidized bed reactor contains the catalyst as a sand like fine granulate on a perforated plate. The gaseous reactants enter the reactor from the bottom and blow the catalyst upwards like a fountain. The catalyst facilitates the reaction and acts also as a heat transfer agent.

Three phase reactions between a gas, a liquid, and a solid catalyst can be investigated in bubble columns with a suspended catalyst or other types of suspension reactors. The disadvantages are the same as for two phase suspension reactors. Therefore trickle bed reactors are used more often. In a trickle bed reactor the catalyst is present as a fixed bed in a packed column. The liquid reactant is sprayed over the catalyst from above and trickles through the catalyst bed. The gaseous reactant enters the reactor at the top or the bottom. It must diffuse through the liquid film on the surface of the catalyst, where it reacts. The gas is commonly recycled.

10.4.2. EXAMPLES OF DIFFERENTIAL REACTORS

In its simplest form a differential reactor consists of a tube containing a small catalyst layer, through which the reaction mixture passes once. The length of the catalyst bed is such that the contact times are very short and that the conversions are kept low ($<5\%$). It is therefore justified to assume that the concentrations of the reactants and hence the reaction rates are constant at every point in the catalyst bed. By analyzing the inlet and outlet streams of the reactor it is possible to calculate the reaction rates.

In a recycle differential reactor the bulk of the gases is recycled outside the heated reactor. A small amount of the products is removed and replaced by an equivalent amount of fresh reactants in each pass.

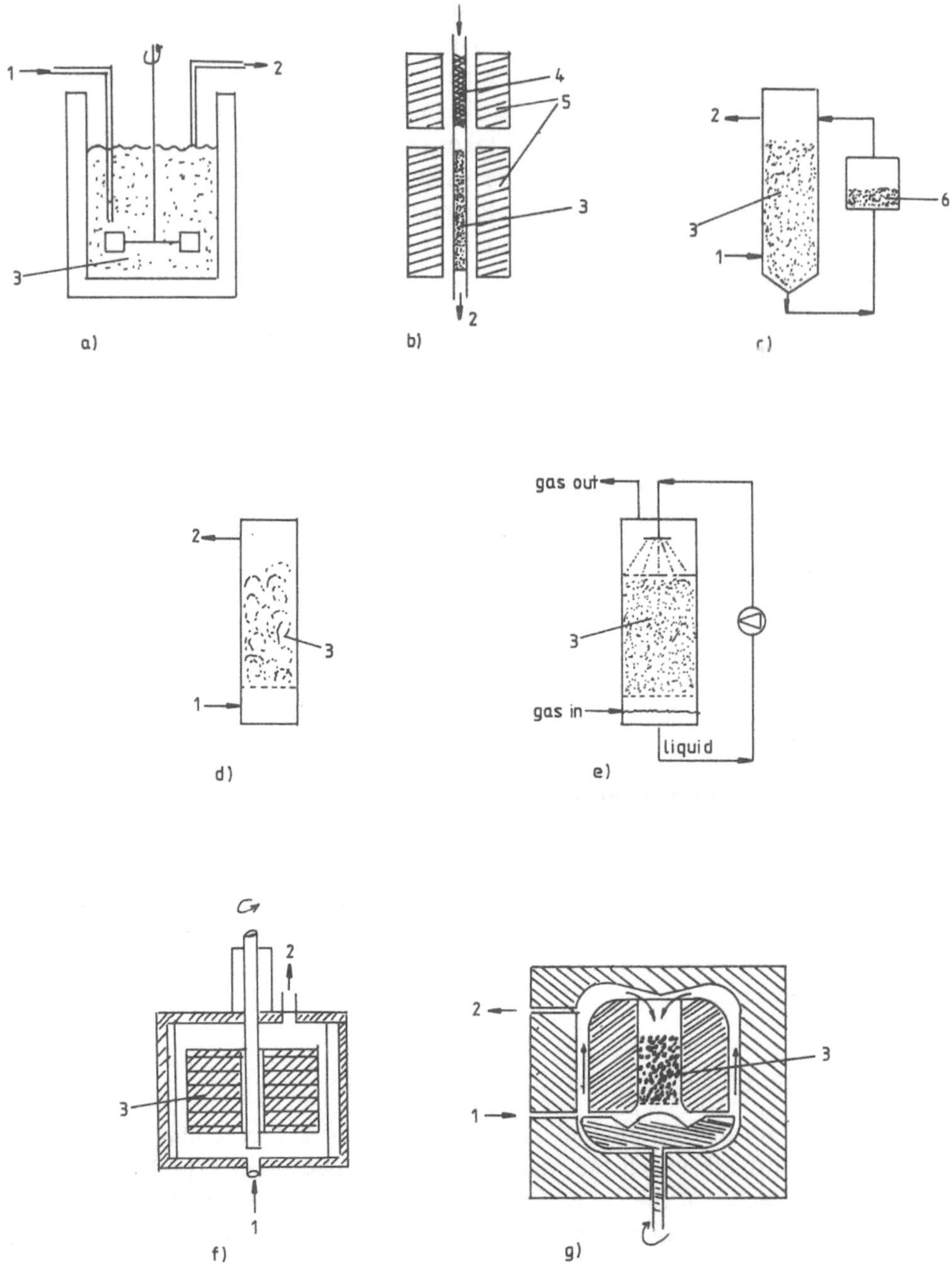

Fig. 10-2. Different types of reactors for catalytic reactions: (a) stirred tank reactor for continuous operation, (b) fixed-bed tubular reactor, (c) moving bed reactor, (d) fluidized bed reactor, (e) trickle bed reactor, (f) rotating basked ('Carberry') reactor, (g) continuous stirred-gas ('Berty') reactor. 1 = reactants in; 2 = products out; 3 = catalyst; 4 = preheater; 5 = furnace; 6 = catalyst regeneration.

In the case of the Carberry reactor the catalyst is placed in a rotating basket, which rotates at high speed in the reaction mixture. It is thereby possible to achieve gradient-free conditions. However, the temperature and the speed of the gas over the catalyst is very difficult to determine, and only catalyst particles with a certain minimum size can be used in the basket.

The Berty continuously stirred gas reactor is frequently used in industrial laboratories. An impeller pulls the reaction mixture through the catalyst bed in the central tube and passes it around through the external opening to the top of the catalyst bed. A small amount of the mixture is withdrawn through a lateral outlet and replaced by an equivalent amount of fresh feed. The reactor can be operated under isothermal conditions, the temperature can be measured directly, and the ratio of free volume to catalyst volume is small.

A single pellet reactor can be used to test individual catalyst pellets. Since each pellet, however, is an individual entity and may differ from other pellets it is necessary to investigate a large number of pellets to arrive at representative conclusions.

It is important to know whether an observed reaction rate is determined by chemical or physical phenomena, such as mass or heat transfer. To avoid physical limitations it is advisable to follow certain rules:

- Don't make the reaction too fast (to avoid diffusion limitation).
- Maintain some turbulence in the gas stream (Reynolds number > 30).
- The ratio of the length of the reactor to the length of a catalyst pellet should be between 30 and 100.
- The ratio of the reactor diameter to the diameter of a catalyst pellet should be between 5 and 8.

10.5. Preparation of Homogeneous Catalysts

Homogeneous catalysts are usually well-defined chemical compounds, which are prepared by normal synthetic methods. They are added to the reactor which contains the reactants as a solution. A simple example is for instance the acid-catalyzed esterification of a carbonic acid with an alcohol. The catalyst is concentrated sulfuric acid, which is simply added dropwise to the stirred solution of the two reactants.

A more complex example is the famous catalyst $HRhCO(PPh_3)_3$. It is prepared from rhodium chloride in a two-step synthesis:

$$RhCl_3 \cdot 3\,H_2O + 2\,PPh_3 + CH_2O \rightarrow ClRhCO(PPh_3)_2 + 2\,HCl + 3\,H_2O$$

$$ClRhCO(PPh_3)_2 + PPh_3 + H_2 \rightarrow HRhCO(PPh_3)_3 + HCl$$

It is isolated in crystalline form and used as a very active catalyst for hydrogenation, hydroformylation, and similar reactions.

Sometimes the catalytically active compounds are prepared *in-situ*, that means in the same reaction vessel in which the catalytic reaction takes place. In such a case a suitable catalyst precursor and the reactants are heated together to the reaction temperature. Under these conditions the precursor is converted into a new compound, which acts as the actual catalyst. A well-known example of a catalyst precursor is dicobalt octacarbonyl. When heated in the presence of a solvent, carbon monoxide, and hydrogen, it is converted into $HCo(CO)_4$, which is an active catalyst for the hydroformylation of olefins. Sometimes other cobalt compounds, for example cobalt acetate, are used as catalyst precursors. Under suitable conditions — high temperature, high CO/H_2 pressure, solvent, and activator — cobalt salts are directly converted to $HCo(CO)_4$.

$$Co_2(CO)_8 + H_2 \rightleftharpoons 2\ HCo(CO)_4$$

$$2\ Co(OAc)_2 + 8\ CO + 3\ H_2 \rightarrow 2\ HCo(CO)_4 + 4\ HOAc$$

10.6. Characterization of Homogeneous Catalysts

If the homogeneous catalysts are stable compounds, they can be characterized by the methods common in inorganic or organometallic chemistry. However, many active catalysts are formed only under the conditions of the catalytic reaction, which often means high temperature and high pressure. Under these extreme conditions special *in situ* techniques must be used.

Composition: The composition of the pure catalyst is determined by elemental analysis. There are no general methods to determine the chemical composition of an *in situ* catalyst.

Structure: The molecular structure of a pure compound catalyst is determined by X-ray diffraction on a single crystal or by neutron diffraction. Information on the molecular structure of catalysts in solution can be obtained by spectroscopic methods, for example NMR or IR (especially for metal carbonyls).

Catalytic cycle: During the catalytic cycle the starting catalyst is converted into other species, which mediate certain conversions of the substrates. At the end of one catalytic cycle usually one molecule of the reactant is converted into products and the active catalyst is regenerated. The intermediates in such a catalytic cycle are present only in small concentrations. Therefore they often cannot be detected directly. In this case kinetic data must provide indirect evidence of the intermediates of the cycle. Nevertheless *in situ* spectroscopy (IR, UV, NMR, ESR, etc.) is a useful tool to study the mechanisms of catalytic processes under their original conditions. Low-temperature and matrix isolation spectroscopy analyze 'frozen species'. Since many reactions are very slow at low temperature, reactive intermediates can be detected, which are undetectable at temperature at or above room temperature. Kinetic investigations can verify or rule out suggested mechanisms.

Possible intermediates can be prepared separately and tested under the original conditions.

10.7. Testing of Homogeneous Catalysts

The quality of homogeneous catalysts depends on a number of factors, like activity, selectivity, stability, etc. For expensive noble metal catalysts the quantitative recovery is also an important criterion. The purpose of the laboratory experiments is often to find the best reactor for a certain reaction. The standard types of reactors are considered first:

The *stirred tank reactor* is the most common type. Here the reactants and products are completely mixed. Their concentration is the same everywhere in the reactor (= ideal backmixing). When no substrates are added or removed during the reaction (= batchwise operation) the concentration of the reactants and products change with time according to the rate law of the reaction. When equimolar amounts of reactants and products are continuously added and removed, respectively, (= continuous operation), the concentrations of all substrates stay constant with time. Under adiabatic operating conditions the temperature is constant in the case of continuous operation but not in batchwise operation.

In a *tubular reactor* ideally no backmixing takes place as long as the flow is viscous (laminar) and not turbulent. The concentrations of the reactants and products change along the reactor tube, as does the temperature under adiabatic operation. At any given point in the tube the concentrations are constant with time.

An intermediate case is a *cascade of several stirred tank reactors* which are operated continuously and in series. Here each reactor has the characteristics of a stirred tank. The more stirred tanks are added to the cascade the more the overall behavior approaches that of a tube.

Very often one reaction component is a gas. Thus the problem arises of bringing the gas into contact with the reactants in the liquid phase. The type of reactor to be chosen depends, among others, on the following criteria:

- The *interfacial area* determines the rate of mass transfer between gas and liquid phases. Better dissipation of the gas increases the contact area, but requires more energy.
- The *gas : liquid ratio* may vary as much as from 0.1 : 1 to 2000 : 1 depending on the application. This requires different reactor designs.
- The selectivity of the reaction is often determined by the amount of *backmixing* between the products and reactants and the *residence time.*
- The heat of reaction of exothermic processes must be removed, that of endothermic reaction provided. High turbulence in the reactor usually improves *heat transfer*, but uses energy and may lower the selectivity. To find the ideal reactor for a certain reaction is something of an art and requires experience.

We must refer to the literature for further details, which would be beyond the scope of this chapter.

References

Szabo, Z. G. and Kallo, D.: *Contact Catalysis*, Vol. 2, Elsevier, Amsterdam, 1976.

Thomson, S. J. and Webb, G.: *Heterogeneous Catalysis*, Oliver & Boyd, Edinburgh, 1968.

Trimm, D. L.: *Design of Industrial Catalysts*, Elsevier, Amsterdam, 1980.

Thomas, J. M. and Thomas, W. J.: *Introduction to the Principles of Heterogeneous Catalysis*, Academic Press, New York, 1967.

Anderson, R. B.: *Experimental Methods in Catalytic Research*, Academic Press, New York, 1968.

Stiles, A. B.: *Catalyst Manufacture: Laboratory and Commercial Preparations*, M. Dekker, New York, 1983.

Fulton, J. W.: 'Building the mathematical model of the catalyst and the reactor', *Chem. Engn.*, Febr. 1986. 'Selecting the catalyst configuration', *Chem. Engn.*, May 12, 1986; 'Making the Catalyst', *Chem. Engn.*, July 7, 1986; 'Testing the catalyst', *Chem. Engn.*, Oct. 13, 1986; 'Evaluating catalyst economics', *Chem. Engn.*, Jan. 19, 1987.

Kelly, M. J.: *CHEMTECH*, Jan. 1987, p. 31; Febr. 1987, p. 98, March 1987, p. 170; April 1987, p. 232.

Krabetz, R. and Mross, W. D.: *Katalyse, heterogene, und Katalysatoren*, Ullmann Bd. 13, 517; Verlag Chemie (in German).

Hölderich, W., Mross, W. D., and Gallei, E.: 'Methods of characterisation of catalysts', *Arabian J. Sci. Engn.* **10**, 407 (1985); 'Reactors and Apparatus for Heterogeneous Catalysis', *Arabian J. Sci. Engn.* **11**, 33 (1986).

Index

MIX
Papier aus verantwortungsvollen Quellen
Paper from responsible sources
FSC® C105338

If you have any concerns about our products,
you can contact us on
ProductSafety@springernature.com

In case Publisher is established outside the EU,
the EU authorized representative is:
Springer Nature Customer Service Center GmbH
Europaplatz 3, 69115 Heidelberg, Germany

Printed by Libri Plureos GmbH
in Hamburg, Germany